T0391082

Implications of ICT for Islamic Finance and Economics

Yilmaz Bayar
Bandırma Onyedi Eylül University, Turkey

Mohamed Sadok Gassouma
University Ez-Zitouna, Tunisia

Vice President of Editorial	Melissa Wagner
Director of Acquisitions	Mikaela Felty
Director of Book Development	Jocelynn Hessler
Production Manager	Mike Brehm
Cover Design	Phillip Shickler

Published in the United States of America by
IGI Global Scientific Publishing
701 East Chocolate Avenue
Hershey, PA, 17033, USA
Tel: 717-533-8845
Fax: 717-533-7115
Website: https://www.igi-global.com E-mail: cust@igi-global.com

Library of Congress Cataloging-in-Publication Data

Names: Bayar, Yilmaz, 1977- editor. | Gassouma, Mohamed Sadok, 1980- editor.
Title: Implications of ICT for Islamic finance and economics / edited by: Yilmaz Bayar, Mohamed Gassouma.
Description: Hershey, PA : IGI Global Scientific Publishing, [2025] | Includes bibliographical references and index. | Summary: "The objective of the book is to investigate the implications of ICT for Islamic finance and economics from a multidisciplinary perspective"-- Provided by publisher.
Identifiers: LCCN 2024060469 (print) | LCCN 2024060470 (ebook) | ISBN 9798369380796 (hardcover) | ISBN 9798369380802 (paperback) | ISBN 9798369380819 (ebook)
Subjects: LCSH: Finance--Religious aspects--Islam. | Information technology--Economic aspects. | Digital communications--Economic aspects. | Finance--Technological innovations. | Economic development--Technological innovations.
Classification: LCC HC79.I55 I5255 2025 (print) | LCC HC79.I55 (ebook) | DDC 303.48/33--dc23/eng/20250331
LC record available at https://lccn.loc.gov/2024060469
LC ebook record available at https://lccn.loc.gov/2024060470
British Cataloguing in Publication Data
A Cataloguing in Publication record for this book is available from the British Library.

Table of Contents

Detailed Table of Contents

In this chapter related to Geoeconomical and Geopolitical (GG), the author uses the Applied Holistic Mathematical Model (AHMM) for GG(AHMM4GG) that can be applied for economy, finance, societal, and business transformations, assessment of GG Risks (GGRisk), GG Problems (GGProblem) and possible eXtremely High Failure Rates (XHFR). These GGRisks can be mitigated by the use of Critical Success Factors (CSFs) (Trad, & Kalpić, 2020a), which can be used in the context of conflictual regions evolution, like the Middle East Area (MEA) and more specifically Lebanon. This chapter is based on a unique mixed research method that is supported by a mainly qualitative research module (Gunasekare, 2015). The AHMM4GG for conflictual and GG Transformation Projects (simply Projects) uses a scripting environment that can be adopted by a Project and for that goal the author uses it to research the The Islamic Finance and Related Technology Strategic Vision in the Lebanese Context (IFSVLC)

Operational efficiency is crucial for the competitiveness and sustainability of banks, particularly in Islamic banking, which adheres to principles such as the prohibition of interest (riba) and asset-backed financing. This study integrates Data Envelopment Analysis (DEA) with Monte Carlo (MC) simulations to evaluate the operational efficiency of Islamic banks. It focuses on three key analyses: assessing Tamweel Bank's efficiency from 2013 to 2023 using historical financial data; generating simulated data with MC simulations to stabilize efficiency estimates; and comparing Tamweel and Alizz Islamic Banks for 2023. Both Constant Returns to Scale (CCR) and Variable Returns to Scale (BCC) DEA models are applied to capture diverse efficiency dimensions. Findings show significant variability in Tamweel Bank's efficiency, peaking in 2019 but declining by 2023. MC simulations provide an aggregated efficiency benchmark, emphasizing the value of combining deterministic and stochastic methods. The comparative analysis highlights an efficiency gap, offering strategic insights for improvement.

Interest-free banking has been steadily gaining a larger share worldwide, introducing alternative financial products to individuals and businesses. However, since all financial intermediaries operate within the same economic system and the core function of intermediation inherently involves earning profits from borrowing and lending activities, it is not possible to consider interest-free banking as entirely separate from conventional banking. In practice, the differences in terminology and theoretical foundations among banking groups often converge significantly. This study presents the performance of participation banks—the practitioners of interest-free banking in Türkiye—over the past 10 years (December 2014–December 2024). Evaluated alongside the development of deposit banks, the performance results of participation banks highlight their expanding role in the sector. While participation banks have outperformed conventional (deposit) banks in terms of balance sheet growth, their financial results and profitability trends have varied across different years.

Fintech is a concept that encompasses new technologies and financial innovations with the objective of optimizing the delivery and utilization of financial services. The term "Islamic fintech" is employed to describe the transformation of financial services and processes through the integration of technological innovations, including blockchain, artificial intelligence, and big data, with the principles of Islamic finance. The advent of Islamic fintech has precipitated a notable surge in demand for Islamic financial products across regions such as the Middle East, Southeast Asia, and Africa. This study assesses the compatibility of fintech with the ethical and operational framework of Islamic finance, with particular emphasis on interest-free banking, risk sharing, and transparency. Furthermore, the study evaluates the potential benefits and obstacles associated with the digitization of the Islamic finance sector, including the sustainable growth and the advancement of innovative solutions. Additionally, the study assesses the impact of regulations and policies in this context.

Digitalization has brought a significant transformation to the Islamic finance sector, enabling it to reach a wider audience. Traditional Islamic finance models were primarily based on physical branches, face-to-face customer interactions, and paper-based transactions. However, with digitalization, banking, investment, and trade processes have accelerated significantly. In particular, internet banking, mobile applications, and blockchain-based solutions have made Sharia-compliant financial transactions more accessible and secure. This transformation has increased interest in Islamic finance while also sparking new discussions on how traditional methods can be sustained in the digital environment. The aim of this study is to examine the role of ICT in Islamic finance from different perspectives and to make future-oriented inferences.

This chapter explores the challenges hindering the advancement of sustainable finance in the Arab region, including regulatory gaps, limited awareness, and restricted access to funding. Using a qualitative approach based on secondary data, the research identifies key barriers such as inconsistent ESG reporting, inadequate financial infrastructure, and perceived investment risks. To foster a more enabling financial ecosystem, the study recommends regulatory reforms, capacity-building initiatives, and awareness campaigns. Additionally, strengthening development banks, promoting public-private partnerships, and leveraging emerging technologies like blockchain can enhance transparency and financial accessibility. Future research should assess these strategies to position the Arab region as a leader in sustainable finance.

This paper aims to assess the importance of adhering to Sharia principles in both conventional and Islamic fundings. We assumed that these principles apply equally to both Islamic and conventional fundings. We also assume that Sharia principles is

Preface

INTRODUCTION

In recent years, the global financial landscape has witnessed a profound transformation, driven in large part by the rapid advancement of information and communication technologies (ICT). These developments have opened new horizons for innovation, efficiency, and inclusivity across all sectors, including Islamic finance and economics. As editors, we recognized the urgency of exploring how Islamic financial institutions, grounded in unique ethical and jurisprudential principles, are adapting to and capitalizing on these technological shifts. This book, *Implications of ICT for Islamic Finance and Economics*, brings together a diverse array of scholars and practitioners who examine the interplay between emerging technologies and the core values of Islamic finance.

The chapters in this volume offer interdisciplinary perspectives that bridge finance, technology, jurisprudence, and strategic governance. From artificial intelligence-powered recommendation systems and automated Shariah auditing tools to geopolitical modeling and performance benchmarking, each contribution reflects a rigorous analysis rooted in both practical relevance and academic integrity. We sought to assemble a body of work that not only captures the current state of the field but also anticipates future trends that will shape Islamic finance in the decades to come. The diversity of geographic and institutional backgrounds represented in these chapters—from Türkiye to Tunisia, France to Oman—further strengthens the global scope and contextual relevance of this volume.

As editors, we are particularly proud of the balance this book strikes between theoretical depth and applied insights. Our contributors demonstrate that Islamic finance is not a static tradition, but a dynamic and evolving system capable of engaging with complex technological transformations while maintaining fidelity to its ethical foundations. In light of the digital era's challenges and opportunities, this book serves as both a scholarly resource and a roadmap for stakeholders—regulators, banks, fintech entrepreneurs, and researchers—committed to fostering innovation without compromising Shariah compliance.

CHAPTER OVERVIEW

Chapter 1: AI-Driven Product Recommendation Systems for Participation Banking: Enhancing Customer Engagement within Islamic Finance Principles

This chapter investigates the potential of artificial intelligence (AI) to elevate customer engagement within participation banking, a core model in Islamic finance. The authors employ machine learning techniques—specifically K-Means clustering and the FP Growth algorithm—to analyze CRM data and develop a personalized product recommendation system that adheres to Shariah principles.

By aligning AI-driven services with Islamic finance ethics, the study presents a framework for tailoring offerings to customer behavior while maintaining compliance with religious mandates. The dual focus on customer satisfaction and religious adherence marks a significant advancement in how participation banks can deliver services in a competitive digital environment.

Findings demonstrate that AI models can boost customer loyalty and operational efficiency in participation banking. The approach also lays the groundwork for sustainable growth, showing how ethical fintech tools can complement financial innovation within Islamic finance frameworks.

Chapter 2: Investigation of Artificial Intelligence Models (AI) in Shariah Auditing of Islamic and Conventional Financing

This chapter explores the use of AI in conducting Shariah audits for both Islamic and conventional banking institutions in Tunisia. The author introduces an automated classification model that integrates jurisprudential principles into AI algorithms to objectively assess the compliance of financial transactions with Islamic law.

Utilizing a logistic regression model, the study identifies key elements of Shariah legitimacy, such as the balance between money and goods, the prohibition of speculative trade, and the principle of profit-and-loss sharing. The AI system's neutrality across institutional types ensures a balanced audit, avoiding biases linked to bank affiliations.

The research underscores the feasibility of real-time Shariah compliance audits through programmed AI models. This innovation not only enhances audit efficiency but also fosters transparency and integrity across Islamic and hybrid financial institutions.

Chapter 3: Geoeconomic and Geopolitical Transformation Projects: The Islamic Finance and Related Technology Strategic Vision in the Lebanese Context (IFSVLC)

In this chapter, the author presents a holistic framework—the Applied Holistic Mathematical Model (AHMM4GG)—to evaluate geopolitical and economic transformation projects within the Lebanese context. Islamic finance and its associated technologies are examined as strategic tools for mitigating geopolitical risk and enabling sustainable development in conflict-prone regions.

The methodology combines qualitative research with advanced modeling techniques to assess risk, identify potential points of failure, and define Critical Success Factors (CSFs) for transformation. Lebanon serves as a case study to explore how Islamic finance can operate within volatile environments and contribute to resilience and reform.

By applying Islamic financial principles alongside strategic planning tools, the chapter demonstrates a unique interdisciplinary approach. It provides valuable insight into how digital Islamic finance can serve as a catalyst for regional development and stability in the Middle East.

Chapter 4: Optimizing Operational Efficiency in Islamic Banking: Integrating Data Envelopment Analysis and Monte Carlo Simulations

This chapter assesses the operational efficiency of Islamic banks using a hybrid model that combines Data Envelopment Analysis (DEA) with Monte Carlo simulations. The focus is on measuring performance under both deterministic and stochastic conditions, using case studies from Tamweel and Alizz Islamic Banks.

Efficiency trends are evaluated over a decade, revealing critical fluctuations and identifying 2019 as a peak year for Tamweel Bank. Monte Carlo simulations are used to generate comparative benchmarks, enhancing the robustness of the findings and offering predictive insight into operational gaps.

The study not only contributes methodological innovation but also offers strategic recommendations for Islamic banks to improve efficiency. By fusing traditional financial analysis with simulation techniques, the chapter sets a new standard for performance assessment in Islamic finance.

Chapter 5: Performance of Participation Banks in Türkiye

This chapter presents a comprehensive performance analysis of participation banks in Türkiye over a ten-year period. Using financial indicators and sector comparisons, the authors trace the evolving role of interest-free banking in a predominantly conventional financial system.

While participation banks demonstrated superior balance sheet growth compared to their conventional counterparts, profitability trends varied, reflecting market complexities and strategic choices. The study highlights both the promise and the limitations of interest-free banking within a shared economic ecosystem.

The authors conclude that despite theoretical and terminological differences, the operational models of Islamic and conventional banks increasingly converge. This finding prompts renewed reflection on the adaptability and strategic positioning of participation banks in global finance.

Chapter 6: The Expansion of Islamic Fintech and the Digital Transformation of Financial Services in the World and Turkey

This chapter explores the rise of Islamic fintech and its transformative impact on global and Turkish financial landscapes. It discusses the integration of digital innovations—such as blockchain, AI, and big data—with Islamic financial principles to deliver transparent, interest-free, and ethically grounded financial services.

Islamic fintech is shown to drive demand for Shariah-compliant products across emerging markets, especially in regions like the Middle East, Southeast Asia, and Africa. The authors evaluate both the opportunities and the constraints facing the sector, including regulatory ambiguity and technological adaptation.

By aligning modern fintech tools with the core values of Islamic finance, the study affirms the sector's potential to foster sustainable, inclusive growth. It also calls for robust regulatory frameworks to support its continued development.

Chapter 7: The Impact of Information and Communication Technologies (ICT) on Islamic Economy and Finance

This chapter analyzes how information and communication technologies (ICT) have reshaped the landscape of Islamic finance. With digital tools accelerating the transition from paper-based to fully digital systems,

Shariah-compliant banking and trade have become more accessible and efficient.

The study details how innovations like mobile apps, internet banking, and blockchain platforms enhance service delivery while maintaining religious observance. It also addresses tensions between modern digital systems and traditional Islamic financial models.

By examining these dynamics, the chapter offers a future-oriented perspective on the digital transformation of Islamic finance. It provides actionable insights for stakeholders seeking to balance innovation with faith-based financial principles.

Chapter 8: The Role of Government in Promoting Sustainable Finance: A Pathway to Achieving the SDGs in Arab Countries

Focusing on the Arab region, this chapter investigates the role of government policy in advancing sustainable finance and achieving the Sustainable Development Goals (SDGs). The analysis identifies key structural and regulatory obstacles that hinder financial innovation and inclusion.

Recommendations include institutional reforms, public-private partnerships, and leveraging emerging technologies to boost transparency and accountability. The use of blockchain, in particular, is highlighted as a tool for enhancing financial system trustworthiness.

The chapter advocates for a proactive role by governments in fostering an enabling ecosystem for sustainable Islamic finance. It offers a roadmap for aligning national financial strategies with global development priorities.

Chapter 9: Quantitative Modelling of Shariah Principles of Islamic Transactions: Empirical Validation in Islamic and Conventional Fundings

In this chapter, the author examines the universal applicability of Shariah principles across both Islamic and conventional banking systems. Through econometric modeling and ratio analysis, the study evaluates adherence to principles such as fairness in trade, labor value, and the prohibition of speculative profit.

Interestingly, findings reveal that conventional banking systems sometimes better uphold certain core Shariah principles than their Islamic counterparts. However, Islamic finance demonstrates superiority in ethical safeguards such as anti-fraud and anti-manipulation mechanisms.

This comparative study challenges assumptions about financial orthodoxy and encourages further refinement of Islamic finance models to ensure holistic adherence to their ethical foundations.

The final chapters underscore critical issues facing Islamic finance in today's interconnected world. From the impact of fintech on financial inclusivity and risk sharing to the role of governments in promoting sustainable finance in line with the UN Sustainable Development Goals, the book broadens its scope to address macro-level concerns. The inclusion of regulatory, policy, and legal perspectives affirms the importance of ecosystem-wide collaboration in building an inclusive, resilient, and ethically grounded financial infrastructure. These insights invite meaningful reflection and action at institutional, national, and international levels.

Moreover, the comparative analyses found in several chapters challenge us to rethink the boundaries and assumptions traditionally placed on Islamic and conventional finance. Whether it is through automated Shariah audits or the modeling of Shariah principles across both systems, these contributions push the frontier of Islamic economic thought and practice. Such inquiries are not merely academic; they speak directly to the operational, reputational, and strategic imperatives of financial institutions operating in Muslim-majority and minority contexts alike.

We extend our gratitude to all contributing authors for their scholarly dedication and intellectual generosity. We also thank our respective institutions for their continued support of interdisciplinary research. It is our hope that this volume inspires further exploration, innovation, and dialogue at the intersection of ICT and Islamic finance. In a rapidly changing world, the ability of Islamic financial systems to remain relevant and impactful will depend not only on technological adoption, but on the enduring commitment to justice, transparency, and ethical stewardship that lie at the heart of the Islamic economic paradigm.

Yilmaz Bayar
Bandirma Onyedi Eylul University, Turkey

Mohamed Sadok Gassouma
University Ez-Zitouna, Tunisia

Chapter 1
AI-Driven Product Recommendation Systems for Participation Banking Enhancing Customer Engagement Within Islamic Finance Principles

Pelin Nazlı Ezber
Turkish Airlines, Turkey

Ersin Namli
https://orcid.org/0000-0001-5980-9152
Istanbul University-Cerrahpasa, Turkey

ABSTRACT

ABSTRACT This study explores the application of artificial intelligence (AI) in participation banking to enhance customer engagement and satisfaction through targeted product recommendations, aligned with the principles of Islamic finance. A random sample of CRM data was analyzed using clustering through the K-Means algorithm, followed by association analysis with the FP Growth algorithm.. This model enables participation banks to provide

DOI: 10.4018/979-8-3693-8079-6.ch001

personalized, compliant product recommendations that increase customer activity and loyalty. Results indicate that machine learning techniques can offer valuable insights into customer behavior patterns, allowing banks to enhance their service offerings while adhering to Islamic finance principles. By supporting customer-centered service delivery, this AI-driven approach contributes to sustainable growth within the participation banking sector.

INTRODUCTION

Participation banking emerged to meet the needs of individuals and businesses seeking interest-free financial solutions, providing an alternative for savers who want to avoid interest-based systems and for business owners seeking ethical financing options. This sector has not only integrated idle resources into the economy but has also expanded banking's scope by diversifying fund collection and distribution through innovative, compliant products. Unlike conventional banks, participation banks operate under stricter limitations due to adherence to Islamic principles. The Participation Banks Association of Turkey and the International Islamic Financial Market (IIFM) establish and publish standardized agreements on Islamic financial instruments, guiding these operations and emphasizing ethical compliance. However, within these frameworks, customer acquisition remains a significant challenge.

Despite extensive customer-focused efforts within bank departments, machine learning offers a powerful and efficient means to enhance prediction and recommendation capabilities, achieving highly accurate outcomes. Machine learning, a branch of artificial intelligence, uses data and algorithms to replicate human learning processes, enabling patterns in customer behavior to be captured and analyzed for future predictions. Maintaining customer engagement requires delivering tailored product recommendations aligned with customer needs, which machine learning facilitates effectively.

This study applies the K-Means clustering algorithm followed by FP Growth association analysis on randomized CRM data in RapidMiner. This approach aims to keep customers active by offering optimized product recommendations within the context of participation banking, contributing to customer satisfaction and loyalty through advanced, AI-supported personalization.

ISLAMIC FINANCE

The Concept of Interest in Islamic Finance

In Islamic finance, instead of interest, the term riba, which is of Arabic origin and means excess or increase, is used. In its simplest form, it can be defined as the excess that occurs in the amount based on time in borrowing transactions. According to the definition of TKBB, interest is the excess or deficiency obtained from money trade without participating in the risk.

People who are accustomed to the conventional banking system and who are not familiar with participation banking think that working with zero interest will result in an environment where investment cannot be made and zero economic growth since profit cannot be obtained. This is a system of thought that completely equates earning with interest. However, in trade, there is return and profit as well as loss, and the participation banking system takes this into consideration and is based on the fair sharing of the output resulting from the transaction according to the risk taken by the parties.

Difference Between Interest and Trade

In the trade of goods & services, it is a requirement of trade and natural for the seller to add his profit to the sale amount. Selling money (and the like) above its value is interest. In other words, it is permissible for a person to sell his product by adding his profit to the cost in exchange for the goods or services he produces, because there is a labor, product or service. The amount added here is profit share and should not be confused with interest. In the simplest terms, interest can be defined as making money out of money. It is interest when the money lent is more than the money borrowed or when money is sold on credit. The biggest benefit of the prohibition of interest is that it encourages people to produce. While trade leads to production, the sale of money by adding profit leads people to production, work and diligence. A dynamic, functioning economy is the most natural outcome of this scenario. In other words, Islamic finance protects not only people but also the economy, society and the environment through socially and environmentally sensitive investments.

Types of Interest

According to the Islamic perspective, money is not wealth; it is a means of acquiring wealth. According to Bukhari, the Prophet Muhammad said, "Gold is exchanged for gold, silver for silver, wheat for wheat, barley for barley, dates for dates, and salt for salt, for like, and in cash. If different types are exchanged, sell as you wish, provided that it is paid in cash." With the development of the economy, this hadith has become a condition of cash payment by the scholars of Islamic jurisprudence in the exchange of "values that replace money."

Jahiliyya Interest

The increase in the payment period given to the buyer on the condition that the debt will increase if the specified term expires and the debt is not paid is Ignorance Interest. In the modern world, its equivalent is literally usury.

Excess Interest

When different types of products are exchanged, it is trade for the seller to make his profit, but when one party exchanges values of the same type, it is excess interest.

Nesîe Interest

If the exchange of the same values is in question during shopping and the excess received on the said value in return for deferred payment is Nesîe interest.

Participation Banking

As an interest-free banking model in terms of its basic structure, it is possible to define participation banking as a system in which monetary transactions and movements of goods and services are closely interconnected, each monetary movement necessarily corresponds to a partnership or goods or services; and the income is shared according to the principles of partnership. (Özsoy, 2012) Participation Banks operating in accordance with

Islamic principles often conduct transactions in unorganized money markets using money market instruments developed in accordance with Islamic principles, not only to meet funding needs or evaluate fund surpluses, but also to manage liquidity risks.

The Islamic financial assessment of the time value of money is not related to loanable transactions, but only to deferred exchange. (Shafii et al., 2013) At the same time, Islamic finance aims to protect the excessive exploitation of the lender and accepts the borrower to repay the principal without paying any additional amount as interest. This reflects true charitable loans that support the welfare of society. Considering the poor in society, there is a provision for extending the loan period in case the lender has difficulty in repaying the loan within the agreed period so that the lender can give more opportunities for the continuity of the business. (Hassan et al., 2019)

Participation banking, which differs from traditional banking in many ways, has the following differences:

Figure 1. Comparison of traditional banking and participation banking

Traditional Banking	Difference Type	Participation Banking
Every legal action is taken.	Operation	Islamic finance instruments are used.
Risk comes with a cost.	Risk	Risk requires sharing.
Money has a time value and this value always increases in favor of the lender.	Partnership	Money has no time value. Since it is obtained through partnership, losing value is as normal as gaining value. Also both partners are affected by this situation in proportion to their contribution.
Income is obtained through interest.	Interest	Income is obtained through profit sharing during the sale of a good or service.
Funds are provided through the credit (lending) method.	Fund Usage Tools	Funds are used through trade or partnership methods.
Cash funds are used.	Fund Usage Method	Cash funds are not used, products or services are always funded.

The basis of the participation philosophy lies in the verse in the Surah An-Najm, which states that man exists only through his labor. Participation banking is a dynamic structure and should be in constant development without deviating from its own principles. When we look at the participation philosophy, we see that man's participation in life, work and the universe takes place. One of the most fundamental differences that distinguishes participation banks from conventional banks is that in partnerships, in accordance with this philosophy, the parties share both profit and loss together. With profit-loss partnership, the profit of the parties is shared in proportion to the input they put forward. In this way, complete justice is achieved. Income equality is nurtured among the society.

All the principles of participation banking are intertwined and in harmony with each other. The details are as follows:

The Principle of Justice

Uphold justice for Allah, be witnesses to justice, and do not let your hatred towards a community lead you to injustice. Act justly… (Surah Al-Maidah, 8) clearly states that justice should be practiced in every aspect of life. Since the source of Islamic finance is directly the Quran, justice is also the most fundamental principle of participation banking.

Tawhid Principle

Tawhid is the state of being one and united in equality. It is necessary to provide mutual benefits. Another main element that should be mentioned about the principle of tawhid is contracts. Contracts are important in every field in Islam, as well as in the field of finance. The process and result of the work done with contracts are transparent, transactions are recorded and accountable. In short; contracts are indispensable for achieving a sustainable Islamic finance system.

Mizan Principle

The Mizan principle is based on the verse "Do not exceed the limits in the Balance…" (Surah Rahman, 8) and can be interpreted from different perspectives. The first of these interpretations is that it is necessary to act honestly in trade, considering that mizan means scale in Arabic. Scales also symbolize equality and balance.

Interest Prohibition Principle

The way for trade to function and for the balance between production and consumption to be achieved is through production. Every production contains value and labor. Earning without labor, consuming without producing are scenarios that participation banking tries to prevent. Interest is another name for earning money without labor and is strictly prohibited.

Risk Partnership Principle

With the risk partnership principle, which is one of the basic principles of interest-free finance, the return in partnerships is not fixed from the beginning but is shared with the outcome of the transaction being known. In this way, justice is ensured in the transaction made because it is not guaranteed under any circumstances that every commercial enterprise will result in profit. While this profit sharing in conventional banking is entirely aimed at protecting the lender, that is, the capitalist, participation banking tries to provide equal benefits to the parties.

Fund Collection Methods

Since it is very difficult for individual individuals and institutions to access financing resources, collecting the capital in the hands of savers and lending it to individuals and companies in need is a pattern that has continued from the past to the present, and intermediary institutions are needed for this. (Hassan et al., 2019) defines participation banks as important structures that fulfill this role by taking Islamic values into account and fulfilling a very important function in terms of interest-free finance.

There are 2 types of fundraising tools: current accounts and participation accounts. These tools are detailed below.

Current Account

A current account in a participation bank is a type of account used for financial transactions such as making transactions, depositing and withdrawing money, and making payments to customers. Current accounts are one of the most common services offered by participation banks to their customers.

Figure 2. Current account operation

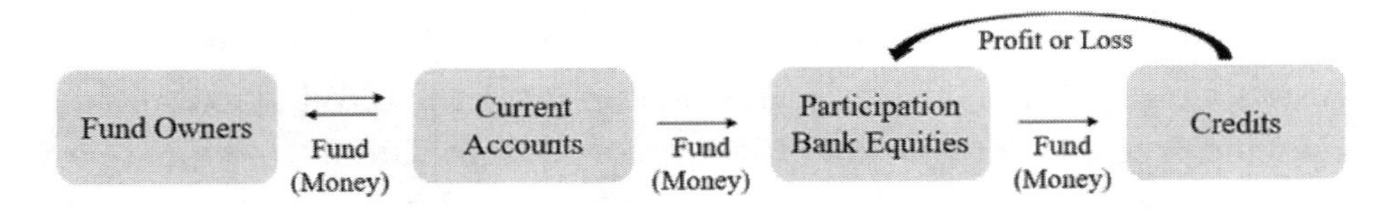

Participation Accounts

Participation accounts in participation banks are an account type based on interest-free banking principles and the principle of profit & loss sharing. This account allows customers to manage their funds interest-free and earn profit shares. The participation account aims to evaluate the funds that customers entrust to the bank through halal investments and commercial activities. This account type is based on interest-free financing and profit sharing principles in accordance with Islamic finance principles. The customer shares the profit & loss of the bank and the profit & loss is shared at the rate specified at the beginning. This rate is usually 5% customer 95% bank or 10% customer 90% bank. In this account type, the customer is not promised to be paid profit at the end of the term, and there is no guarantee that the principal he initially deposited will be paid, because the bank may have made a loss as well as a profit.

Figure 3. Participant account operation

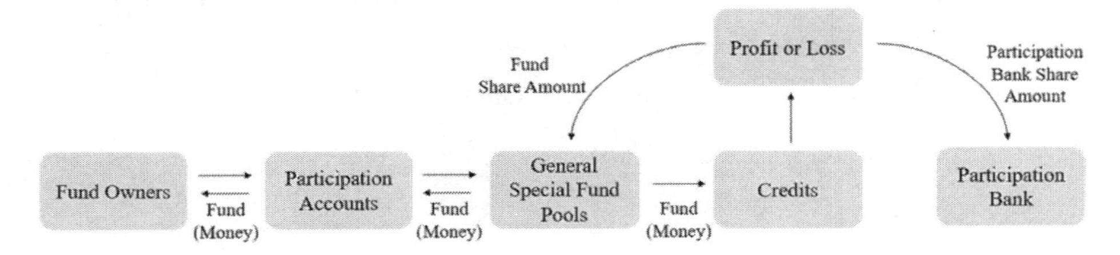

Fund Utilization Methods

Before modern banking, Muslims in need of cash used different solutions in the borrowing process. Before the development of Islamic finance and the emergence of participation banks, due to the lack of interest-free financing methods in accordance with Islam, Muslims found some alternative ways to meet their cash needs. Mutual assistance among those in need is a fundamental value in Muslim communities among individuals in need of cash, family, relatives, neighbors and other members of society, and people have mostly received support based on the principles of cooperation and solidarity in periods when there were no interest-free banks. This included aid and lending methods such as zakat, charity or karz al-hasan

In the modern period, participation banks and Islamic financial institutions have solved this problem by offering interest-free financing opportunities to Muslims in need of cash. These institutions meet the cash needs of Muslim individuals and businesses using various fund usage methods based on interest-free financing principles. The funds collected in participation banking are distributed in accordance with Islamic principles and the Turkish legal system. Principles such as interest-free, risk partnership and justice are also taken into consideration when using funds. According to the Regulation on Bank Credit Transactions of the Banking Law, corporate finance support is the process of indebting the business on the condition that the price of all kinds of goods, securities, real estate, rights and services required by the business is paid to the seller within the scope of the contract to be concluded between the participation bank and the person who will use the fund. Individual finance support is defined as the indebting of the buyer on the condition that the price of goods or services purchased directly from sellers by real person buyers for individual needs is paid to the seller by the participation bank. (Banking Law No. 5411, Regulation on Credit Transactions of Banks, Article 19, Paragraph 2)

Since the system also operates entirely with commercial logic, if there is no goods or services, the participation bank cannot provide funds. In short, the goods needed by the customer are purchased instead, profit is added and sold to the customer on credit.

Islamic markets offer different tools such as sales, trade and financing to satisfy fund providers and users in various ways. The basic fundraising methods it uses are as follows: cash purchase and deferred sale (murabaha), profit sharing (mudaraba), leasing (icare), partnership (musharakah) and deferred sale (salaam), specialized deferred sale (istisna) and sukuk. These instruments serve as the building blocks for the development of a wide range of more complex financial instruments and are seen as a great potential for financial innovation and expansion in Islamic financial markets.

Murabaha

Murabaha comes from the Arabic word "ribh" meaning profit and gain. While the murabaha system is logically clear and fixed, the ways it works can vary between institutions. Murabaha can be briefly defined as selling a good or service with a profit added to it. (Özsoy, 2012)

Mudaraba (Community of Interest)

Mudaraba can be defined in its simplest form as a partnership between labor and capital. In fact, mudaraba has been resorted to in many areas of society since ancient times without being fully aware of it. In this partnership, one party has money but is not in a position to work or has no idea. The other party has knowledge about the situation that is the subject of the partnership, is in a position to work, but does not have capital. In such cases, these two parties come together. One puts forth money and the other puts forth labor and enters into a partnership.

Musharakah (Partnership)

Musharaka is one of the types of partnerships made according to the profit-loss sharing principle. The word also means partnership in Arabic. In other words, according to the profit-loss sharing principle, the loss is shared in proportion to the capital contributed. The profit sharing ratio is determined by everyone's consent when establishing a partnership between the partners.

Salam

Unlike other fund usage methods, Salam is based on the principle of first delivering the goods and then paying for them. It has been used frequently since the early days of Islam. The agricultural sector is especially the most basic area of use.

Exception

In the exception, which is quite similar to Salam in terms of operation, payment is made before the goods are received, as in Salam. The most important feature that distinguishes Istisna from Salam is that while in Salam, the product must be standard, in Exception, the product must be unique. A designer shoe, bag or a specially made wardrobe may be among the products subject to the exception.

Sukuk

The term sukuk, which is similar to the financial certificates in conventional banks in participation banks, comes from the Arabic root saqq, meaning certificate. The plural form of this word, sukuk, means certificates. It can be defined as an asset-backed security in its simplest form.

Murabaha Sukuk

Murabaha sukuk are securities that provide fixed rate of return to the properties of the co-owners under the murabaha contract. (Pireh, 2008)

Murabaha sukuk is a financial instrument issued in accordance with the principles of Islamic finance. Sukuk is also called Islamic bonds or debt instruments and is used to meet financing needs in accordance with Islam.

Ijarah (Leasing)

Ijarah means renting in Arabic. The method called leasing in conventional banks is quite similar to ijarah. In fact, its basic principle is the same. In participation banking, ijarah can be summarized as the renting of products that are in accordance with Islamic law to customers in order to obtain funds after the bank has acquired them in accordance with Islamic principles.

METHODOLOGY

Recommendation Systems

In today's world where there is an abundance of information and options, it is becoming increasingly difficult for users to find the product, service or information they are looking for. Recommendation systems are a technological tool that emerged to solve this problem and facilitate information discovery by offering personalized suggestions to users. In other words, recommendation systems are systems that offer product, service or content recommendations to users based on their personal preferences, past behaviors and demographic characteristics. These systems use various methods to determine users' interests and needs and try to offer the most appropriate recommendations in light of this information. When these systems first emerged in the 1990s, they used very simple algorithms. Over time, with the development of technologies such as machine learning and artificial intelligence, recommendation systems have become more complex and sophisticated.

If we look at the important milestones in the development of recommendation systems, the 1990s can be positioned as the period when basic algorithms such as collaborative filtering and content-based filtering were developed, the 2000s when machine learning techniques were applied to recommendation systems, and the 2010s when deep learning and artificial intelligence began to be used in recommendation systems.

Recommender systems are essential in managing information overload by filtering vast quantities of data to provide users with tailored content based on their preferences. Broadly, these systems are divided into three main types: content-based, collaborative filtering, and hybrid systems. Content-based systems focus on recommending items similar to what a user has liked in the past, based on item attributes, while collaborative filtering relies on preferences expressed by users with similar tastes. Hybrid approaches combine both methods to leverage their strengths and mitigate limitations like the "cold start" problem, where new users or items lack sufficient data for accurate recommendations (Roy & Dutta, 2022; Raza & Ding, 2021).

Methods Used

Figure 4. Types of recommendation systems (Roy & Dutta, 2022)

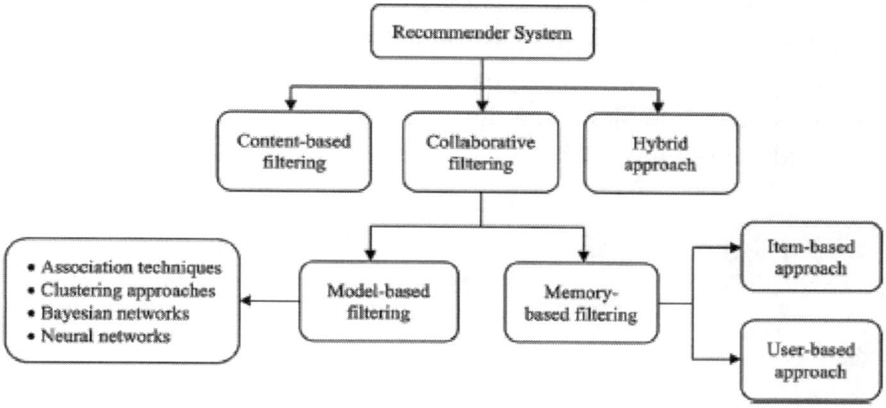

Recommender System

Content-based filtering

Collaborative filtering

Hybrid approach

Model-based filtering

Memory-based filtering

- Association techniques
- Clustering approaches
- Bayesian networks
- Neural networks

Item-based approach

User-based approach

Content-Based Filtering Method

Content-Based Filtering methods are performed by filtering the product that is the subject of the recommendation system according to its content. The criteria determined during filtering may vary depending on the product. The basic logic in the operation of such systems is to collect data about the content, cluster this data and then make a recommendation by querying the relationship with these clusters during the recommendation phase. Weight vectors are taken into account as a measure when calculating the similarities between the contents.

The content-based filtering method has the advantage of offering personalized recommendations based on the user's past preferences and the characteristics of the content, thus increasing user satisfaction by allowing users to access content that is more suitable for their interests (Roy & Dutta, 2022; Raza & Ding, 2021).

Collaborative Filtering Methods

Collaborative approaches take advantage of the similarity measure between users; therefore, regardless of which collaborative method is used, the basic logic is based on clustering. This technique starts by finding a group or collection of users B whose preferences, likes and dislikes are similar to those of user A. New items that are liked by most users in B are then recommended to user A.

Figure 5. Process of collaborative Filtering (Zhou, et al, 2023)

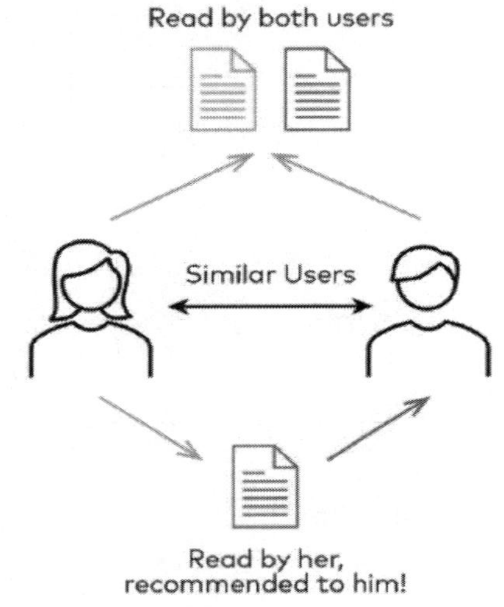

Hybrid Filtering Methods

Hybrid filtering is a filtering method that combines two or more techniques used together to address the limitations of individual recommendation techniques. Combining different techniques can be accomplished in a variety of ways. A hybrid algorithm can combine the results from separate techniques or use content-based filtering in a collaborative method. It can also use a collaborative filtering technique in a content-based method. This hybrid combination of different techniques often results in increased performance and increased accuracy in many recommendation applications.

Hybrid filtering, which is usually implemented as a two-stage process, first generates potential recommendations using content-based information such as the user's preferences and profile information. These recommendations are then combined with collaborative filtering information such as the preferences of similar users to obtain more precise and personalized recommendations (Roy & Dutta, 2022; Raza & Ding, 2021).

Machine Learning

Machine learning (ML) is a field focused on developing algorithms that enable computers to learn patterns from data without being explicitly programmed (Güler et al, 2024; Güler & Namlı, 2024). At its core, ML transforms data (experience) into predictive or decision-making capabilities, often through algorithms that adaptively improve performance on tasks as they process more data. The main categories of ML include supervised learning—where algorithms are trained on labeled data to predict outcomes, and unsupervised learning—which identifies patterns in unlabeled data, such as clustering similar items. Reinforcement learning is another approach where an agent learns by receiving feedback from actions taken in a given environment. ML models benefit from methods like stochastic gradient descent for optimization, and advanced techniques such as neural networks and support vector machines are commonly applied in domains from image recognition to natural language processing (Shalev-Shwartz & Ben-David, 2014; Han & Kamber, 2006; Witten et al., 2017; Ünlü, 2022; Güler et al, 2024).

Clustering

Clustering is an unsupervised machine learning technique that groups similar data points based on predefined criteria, enabling patterns and structures to emerge from raw data. Common clustering algorithms include K-means, which partitions data into a predetermined number of clusters, and hierarchical clustering, which builds a tree-like structure to represent nested groupings. Another approach, density-based clustering, identifies clusters based on areas of high data point density, suitable for complex, non-linear shapes. These methods are foundational for tasks such as image segmentation, customer segmentation in marketing, and anomaly detection, offering insights without requiring labeled data (Witten et al., 2017; Ashendena et al., 2021; Kshatri et al., 2023; Ünlü & Xanthopoulos, 2021).

Association Analysis

Association analysis is an important method in data mining and is used to determine the relationships and associated rules between elements in a dataset. It is also commonly known as market basket analysis because it is widely used to understand customer shopping behavior and identify critical relationships, especially in the retail sector. The reason is that it is based on the principle that items that appear together tend to occur together frequently. This means that customers who purchase one product may also tend to purchase other specific products. Thus, association analysis has become a method used in various industries and business areas. It is widely used, especially in the retail sector, to understand customer purchasing habits and develop recommendation systems. However, web mining is also used in other areas such as telecommunications, healthcare, financial services, and marketing.

Association analysis is a data mining technique used to uncover relationships among items within large datasets, most commonly in transactional data. The method involves identifying patterns of co-occurrence, typically represented as "if-then" association rules, such as "If Item A is bought, then Item B is likely to be bought." This is extensively used in market basket analysis, where retailers analyze customer purchasing behaviors to determine frequently bought item pairs. Key metrics in association analysis include support (the frequency of item sets) and confidence (the likelihood that one item set predicts another), helping refine the quality of extracted rules. The technique has applications in various domains beyond retail, such as healthcare and finance, where identifying associative patterns can drive strategic decisions (Nanopoulos et al., 2007; Zhou & Yau, 2007).

FP-Growth Method

The FP-Growth (Frequent Pattern Growth) algorithm is an efficient method for association rule mining that builds a compressed representation of the dataset known as an FP-tree. Unlike traditional methods like the Apriori algorithm, FP-Growth avoids generating candidate sets by compressing frequent item sets into the tree structure, which enables faster pattern discovery, especially with large datasets. It operates by first scanning the database to identify frequent items, organizing them into a prefix tree, and then recursively exploring conditional trees for each item to mine frequent patterns without scanning the database multiple times. This method is particularly advantageous in handling large datasets where computational efficiency is crucial (Han et al., 2000; Sarath & Ravi, 2013).

The algorithm operates in two main steps:

Constructing the FP-tree: First, FP-Growth scans the database to identify and order frequent items in descending frequency. The database is then compressed into an FP-tree, where nodes represent item occurrences, and branches represent item sequences in transactions. This structure allows the dataset to be efficiently stored without redundant scanning (Sarath & Ravi, 2013; Nanopoulos et al., 2007).

Mining frequent patterns: After building the FP-tree, the algorithm recursively mines frequent patterns by constructing conditional FP-trees for each frequent item, focusing only on those transactions that contain the item being analyzed. This process is highly efficient, especially in sparse datasets, because it reduces the number of database scans needed to identify patterns (Butryna et al., 2021; Zhou & Yau, 2007).

FP-Growth is particularly suited for large-scale data mining tasks because it reduces computational complexity and improves memory efficiency compared to traditional methods. It has been successfully applied in various domains, including retail for cross-selling strategies, financial market predictions, and healthcare analytics, where quick and accurate association pattern discovery is critical (Srivastava et al., 2024; Zhou & Yau, 2007).

The values interpreted in FP Growth Outputs are as follows:

- Support (Support Value): The probability of a product/relationship being seen in all events.
  N: Total number of purchases
  Support (X -> Y) = Frequency(X, Y) / N
- Confidence (Confidence Value): The probability of a customer who buys product X to buy product Y.
  Confidence (X -> Y) = Frequency(X, Y) / Frequency(X)
- Lift: Indicates how many times the occurrence of event X increases the probability of event Y.

$$Lift = Support(X,Y) / (Support(X) \times Support(Y))$$

PURPOSE OF THE STUDY

Today, in participation banks, which have become the preference of customers due to interest sensitivity, it is important to create a product and recommend this product to the most suitable customer by adding a series of restrictions in accordance with the principles of Islamic Finance, in addition to the restrictions coming from traditional banking. In this study, classification, clustering and basket analysis were carried out under the recommendation systems, which are the most powerful branches of machine learning, and the prediction of the potential customers of the bank and the most suitable product for each customer was studied. In customer prediction, the system searches for similarities between the data of a random person and the basic data of the participation bank test environment customers and questions the possibility of being included in any cluster. According to this query, it gives the output as to whether the person is a potential customer. In the prediction of the most suitable product to be recommended to the customer, after grouping the customers according to the products they buy and their demographic data, it is aimed to first assign a customer who enters the system to the group with the customers who are most similar to him, and then compare the list of products frequently purchased by the customers in this group with the products that the new customer has used so far, and recommend the most suitable product to him without spending any effort on marketing irrelevant products.

DATASET

The data set test environment consists of random data, it is not related to any institution. In the data set of 1000 people, there are 20 personal data for each customer and 18 usage information for the product group.

Personal data is textual, product usage information is coded with 0 and 1 according to usage and non-usage status. To give a few examples:

Personal information obtained from random test customers of X bank used in the application is as follows:

- Age
  There are 1000 customers with a minimum age of 14 and a maximum age of 81.
- Age Range
  Customers' ages are divided into 6 groups as 0-20, 21-30, 31-40, 41-50, 51-60, 61 and above.
- Gender
  There are 2 genders: female and male.
- Place of Birth
  It shows the place of birth in the population information.
- Residence
  It shows the province of residence.
- Does he/she live in one of the 5 largest cities in Turkey?
  According to the latest census, the 5 largest cities in Turkey in terms of population are Istanbul, Ankara, Izmir, Bursa and Antalya. If the customer lives in one of these cities, it is coded as "yes", if he/she lives in one of the other cities, it is coded as "no".
- Does he/she live in the city where he/she was born?
  This is the data obtained by comparing the place of residence and place of birth. It consists of two groups as "yes" and "no". It is assumed that the person left the city where he/she was born for reasons such as education or work, and it constitutes one of the inputs of the model as a determining factor.
  Marital Status
  It is divided into 3 groups as married, single and widowed.
- Number of Children
  The number of children is examined in 6 groups as no, 1 child, 2 children, 3 children, 4 and +4 children, unknown.
- Educational Status
  Educational status is examined in 6 groups as literate, primary school, high school, undergraduate, graduate and doctorate according to the last school graduated.
- Customer Becoming Channel
  The customer becoming channel is divided into two as branch and self-service. While the branch is a physical environment, self-service represents all online channels.
- Customer Becoming Date

- Customer Being Duration
  It shows the number of days until today based on the date of December 9, 2023 when the data was obtained.
- Customer Type
  It is a data obtained from the customer being period and is created by clustering the customer's relationship with the bank in time. It consists of 4 groups as 0-1 year, 1-3 years, 3-5 years and old customer according to the customer being period.
- Home Status
  It is divided into 3 groups as present, absent and unknown according to ownership.
- Car Status
  It is divided into 3 groups as present, absent and unknown according to ownership.
- Working Status
  The working status is divided into 4 groups as unemployed, salaried, employer and retired according to the customer's current information.
- Profession
  There are 16 types of occupation information.
- Value Segment
  It is the value calculated by the bank according to the customer's financial situation and activity in the bank, it is expressed with letters. It represents the customer's material value for the bank. The order of values is A1, A2, B, C, D, respectively; P also indicates passive customers.
- Mobile Last Login Day Difference
  The current date shows the number of days from the date the last mobile login was made until this date, based on December 9, 2023, when the data was obtained.
- Mobile Activity
  Assuming that each customer using mobile banking will log in to the mobile at least once a month for bill payments or salary transactions, customers who have logged in to the mobile banking application in the last 31 days at most are considered mobile active, and previously mobile passive.
- Total Product Usage

It indicates the total number of products used by each customer at the relevant bank. It is data obtained later by examining the individual-based product usage.

- Main Bank?

  It is produced by interpreting the total product usage information showing the number of products used by the customer at the bank. If the customer's product count at the bank is 7 and +7, it is interpreted that the customer is sufficiently active at the bank and the relevant bank is coded as the customer's main bank, however, if the total product usage is below 7, it is assumed that the customer is not highly active and is not the main bank of the customers in this group and coded.

Data regarding Product Usage is coded as 1 if used by the relevant customer and 0 if not. The products queried are as follows:

- PPS (private pension system)
- Automatic Payment Order
- HGS (card pass system)
- Life Insurance
- Debit Card
- Credit Card
- Investment Fund
- Home Insurance
- TCIP (turkish catastrophe insurance pool)
- Car Insurance
- Foreign Exchange Account
- Precious Metal Account
- Mobile Branch Usage
- Participation Account
- Premium Credit Card
- Stock

RESULTS

Clustering was done with 1000 rows of test customer data and the details of the work done are as follows.

General Cluster Outputs

It was observed that the smallest Davies Bouldin ratio was reached with 8 clusters. The details of these 8 clusters are as follows:

Figure 6. Distribution of Customers as a Result of Clustering

Cluster Model	
Cluster 0	174 items
Cluster 1	125 items
Cluster 2	136 items
Cluster 3	103 items
Cluster 4	105 items
Cluster 5	137 items
Cluster 6	96 items
Cluster 7	124 items
Total number of items	**1000**

The first output shows how many customers are assigned to each cluster after dividing the dataset consisting of 1000 people into 8 clusters. According to this distribution, 17.4% of the customers are in cluster_0, 12.5% in cluster_1, 13.6% in cluster_2, 10.3% in cluster_3, 10.5% in cluster_4, 13.7% in cluster_5, 9.6% in cluster_6 and 12.4% in cluster_7.

Heat Map, perhaps one of the best outputs of K-Means, shows the distribution density of the data in each cluster as follows. While the data can be distributed with a density below 100% of the average, a cluster with a density of up to 650% was encountered. Therefore, the lower limit of the Heat Map was determined as -100%, while the upper limit was determined as 650%.

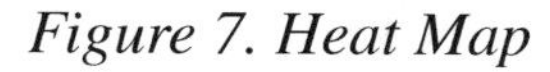

Figure 7. Heat Map

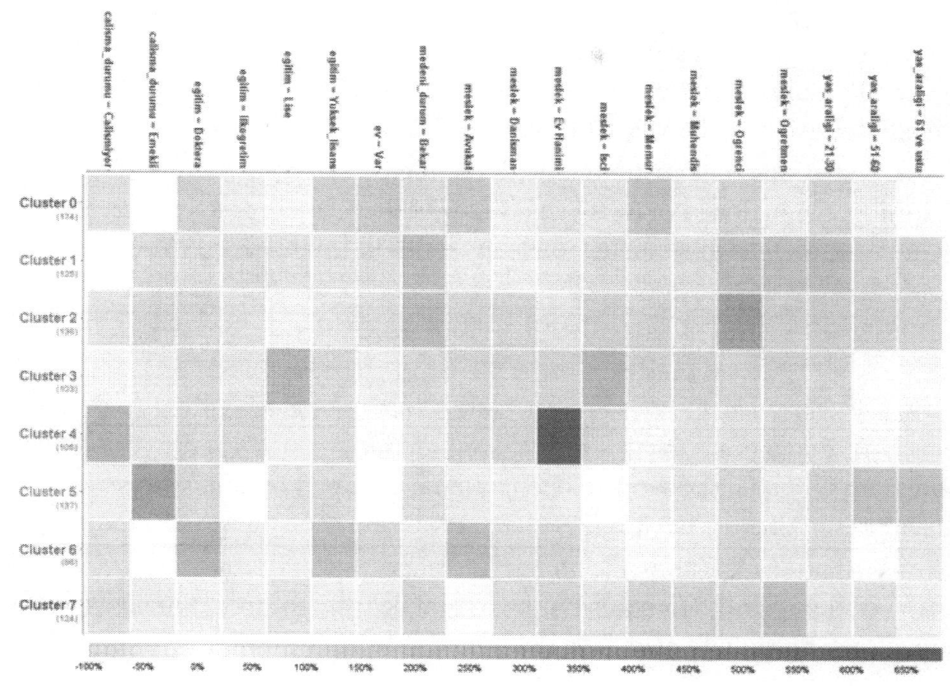

Examining the Created Clusters

Cluster 0: Civil Servant and Lawyer Customers

Features: Customers in this cluster are generally homeowners and work as civil servants or lawyers. Due to these features, it can be interpreted that their level of education is high and their financial situation is good. The cluster consists of individuals who are educated and have a stable income.

Associations: It has been determined that 100% of the customers in this cluster who have a participation account use mobile branches. In addition, it has been observed that customers who have precious metal, participation accounts and foreign exchange accounts also use debit cards.

Cluster 1: Young and Single Students

Features: This cluster consists of student customers who are younger and generally single. The cluster represents student and new graduate groups, mostly individuals between the ages of 21-30.

Associations: All customers with debit cards also have life insurance. In addition, the majority of those who use vehicle financing have life insurance, which shows that this cluster seeks diversity in financial services.

Cluster 2: Young Customers Under Graduate

Characteristics: Customers in Cluster 2 are young individuals with lower levels of education. The highest student density is observed in this cluster and also includes the worker occupational group. These customers mostly have pre-graduate education level and represent the young age group.

Associations: Customers with vehicle financing and participation accounts also use debit cards. In addition, it is seen that 89% of individuals with precious metal accounts also have debit cards.

Cluster 3: Middle-Aged Worker Customers

Characteristics: In this cluster consisting of customers in the middle age group, most of the customers are high school or primary school graduates and work as workers. In this cluster, where married individuals are predominant, some also own houses.

Associations: All customers using housing finance also have life insurance. All those using BES have debit cards and all those with housing finance have TCIP (compulsory earthquake insurance).

Cluster 4: Housewives

Features: In this cluster, where housewife customers are dense, there are mostly primary school graduates, unemployed individuals over the age of 30. Some are high school graduates and their home ownership rates are low.

Associations: 100% of customers with foreign exchange accounts use debit cards. In addition, 94% of those with participation accounts have precious metal accounts, and a strong relationship is observed between these two products.

Cluster 5: Retired Customers

Features: In this cluster, customers over the age of 51 predominate and the majority are retired. Retired customers represent individuals who prefer a quieter lifestyle.

Associations: All those who have insurance have debit cards. The fact that those who use credit cards and HGS also have debit cards and insurance shows that this cluster continues to benefit from financial services even during retirement.

Cluster 6: Highly Educated Customers

Features: Customers included in this cluster have a high level of education. Doctorate and master's degree graduates are densely found in this cluster. Lawyers, engineers and consultants are among the professional groups, and they are usually not homeowners.

Associations: All of those using housing finance also have life insurance. 100% of customers with precious metal accounts use mobile branches, which indicates that this cluster has a high level of financial literacy.

Cluster 7: Young and Educated Employees

Characteristics: This cluster, consisting of customers in the young age group, has an above-average level of education and generally consists of individuals working as teachers, engineers and consultants. The cluster may include customers who are inclined to travel and invest.

Associations: It has been observed that all of those using PPS and vehicle finance have life insurance. The majority of customers with foreign currency accounts use mobile branches, which indicates that it is preferred to facilitate foreign currency transactions.

Association Analysis

After clustering, association analysis was performed with the FP Growth algorithm for both each cluster and each value segment created by the bank. The findings obtained as a result of this analysis study are as follows:

This study performed association analysis with the FP Growth algorithm for each cluster and bank value segment by performing customer segmentation in the banking sector. The analysis details the relationships between customer behavior and product usage of each segment. Trust (conf.), leverage (lift), plausibility (lev.) and conversion (conv.) values show the prominent relationships of each segment and cluster.

FP Growth Applied to Clustered Data

Cluster 0: It was observed that all (100%) customers with participation accounts used mobile branches (conf: 1, lift: 1.07, lev: 0.02, conv: 3.22). The rate of debit card and mobile branch usage of customers with precious metal, foreign exchange and participation accounts increased by 108%.

Cluster 1: 100% of customers with vehicle financing and debit card use life insurance (conf: 1, lift: 1.28, lev: 0.1, conv: 12.74). The probability of vehicle financing and mobile branch usage together is 9% higher.

Cluster 2: All customers using vehicle financing have life insurance (conf: 1, lift: 2.83, lev: 0.2, conv: 26.53). It is an important finding that 97% of customers with participation accounts use debit cards.

Cluster 3: 100% of customers using BES use debit cards (conf: 1, lift: 1.03), but there is no strong connection between these two products.

Cluster 4: All customers with foreign exchange accounts use debit cards (conf: 1, lift: 1.05, conv: 1.52). 94% of those with participation accounts open precious metal accounts.

Cluster 5: All customers with insurance use debit cards (conf: 1, lift: 1.01), but the relationship between them is weak. The use of debit cards and insurance by customers with credit cards and HGS increased by 249%.

Cluster 6: It was observed that all customers using housing finance have life insurance (conf: 1, lift: 2.23, lev: 0.25, conv: 23.74).

Cluster 7: All customers using debit cards and housing finance also have TCIP (conf: 1, lift: 6.53, lev: 0.13, conv: 16.09). There is a strong relationship between these three products.

FP Growth Applied to Bank Value Segments

A1 Segment: 91% of customers using home insurance use Premium credit cards (conf: 0.91, lift: 1.05, lev: 0.02, conv: 0.99). 91% of those with foreign currency accounts use invoice instructions.

A2 Segment: All customers using debit cards and home finance have life insurance (conf: 1, lift: 4.85, lev: 0.13, conv: 12.7). 100% of mobile branch, life insurance and home finance users have DASK (conv: 12.53).

B Segment: All customers using home finance use life insurance (conf: 1, lift: 2.44, lev: 0.13, conv: 27.16). 100% of customers with precious metal accounts and housing finance also have a debit card and DASK (conv: 26.55).

C Segment: It is observed that all customers with life insurance and housing finance have TCIP (conf: 1, lift: 6.96, lev: 0.12, conv: 38.53). 89% of BES users use mobile branch (lift: 1.12, conv: 1.62).

D Segment: 100% of customers with PPS use a debit card (conf: 1, lift: 1.08, conv: 2.47). 97% of those using consumer finance also have a debit card (conv: 1.27).

P Segment: No significant product unity was observed in this segment and customers were evaluated as passive.

DISCUSSION

With the application, first cluster analysis was performed, then association analysis was performed separately on the created clusters. During clustering, the K-Means algorithm was applied, and then association analysis was performed on each cluster with the FP Growth algorithm. In addition, since the segmentation carried out by the bank was a type of clustering study, the FP Growth algorithm was applied to each segment and the results were compared. According to the analysis results;

- Since customers were segmented according to the products they used and their financial savings in the bank in the current system implemented by the bank, the bank could miss the opportunity to sell to potential customers (due to not being active); however, in the clusters created with K-Means, customers were examined in terms of products used and financial savings, and demographic features were also examined to determine the characteristic features of the clusters. In this way, better identification of customer consumption habits was achieved.
- Since much more rules and relationships were detected in the association analyses applied to the clusters created with the K-Means algorithm than in the association analyses applied to the clusters formed by the bank's segmentation, the bank's product sales will be increased by better identification of customer needs.
- It has been shown that by clustering customers in any participation bank, the customers accumulated in each cluster and the most suitable products for the customers, and how the customers in the cluster purchase the products in combinations can be measured.
- It is expected that customer activity and loyalty will increase by recommending the most suitable product to each customer.

The biggest difficulty experienced during the study was that there are very few artificial intelligence-supported studies in the field of Islamic finance in the literature. For this reason, there is no study that includes a product recommendation model that can be cited and the results can be compared. This study aims to offer a different perspective to the literature filled with a traditional perspective towards participation banking.

CONCLUSION AND RECOMMENDATIONS

Today, the vast majority of customers who choose participation banking make their choice with interest sensitivity. As a result of this sensitivity, they do not deserve to encounter a less efficient system in terms of both product and service compared to a bank operating with interest. On the other hand, participation banks can only take action with the extra restrictions determined by both the state and private institutions. Within these restrictions, it is of critical importance for participation banks trying to compete with conventional banks to identify their potential customers, recommend the most suitable product to their customers and continue their work with maximum efficiency. By finding the right customer and recommending the most suitable product to this customer, they will reach maximum efficiency without incurring PR costs. This study proves how beneficial it will be for participation banks to benefit from artificial intelligence-based product recommendation models when recommending products to their customers. In the last leg of the analysis, the results of FP Growth applied to the value segments determined by the bank were brought together with the results obtained from the clustering study. The findings revealed that while the clustering-based approach showed superior performance, there was a significant inequality in the effectiveness of the relationship analysis. This meticulous analysis study, consisting of clustering and association analysis, has provided a detailed understanding of customer behaviors and preferences. It is recommended that applications be made with larger data and various hybrid proposition models in studies to be carried out based on this study. In this way, a different perspective will be brought to the model and the study will be improved. In addition, the inclusion of the reward system in the bank will be effective in increasing the attractiveness of the recommended product.

REFERENCES

Ashendena, S. K., Bartosik, A., Agapow, P.-M., & Semenova, E. (2021). *Introduction to artificial intelligence and machine learning. The Era of Artificial Intelligence, Machine Learning, and Data Science in the Pharmaceutical Industry*. Elsevier.

Butryna, B., Chomiak-Orsa, I., Hauke, K., Pondel, M., & Siennicka, A. (2021). Application of Machine Learning in medical data analysis illustrated with an example of association rules. *Procedia Computer Science, 192*, 3134–3143. DOI: 10.1016/j.procs.2021.09.086

Güler, M., Kabakçı, A., Koç, Ö., Eraslan, E., Derin, K. H., Güler, M., Ünlü, R., Türkan, Y. S., & Namlı, E. (2024). Forecasting of the Unemployment Rate in Turkey: Comparison of the Machine Learning Models. *Sustainability (Basel), 16*(15), 6509. DOI: 10.3390/su16156509

Güler, M., & Namlı, E. (2024). Brain Tumor Detection with Deep Learning Methods' Classifier Optimization Using Medical Images. *Applied Sciences (Basel, Switzerland), 14*(2), 642. DOI: 10.3390/app14020642

Han, J., & Kamber, M. (2006). *Data Mining: Concepts and Techniques*. Morgan Kaufmann.

Han, J., Pei, J., & Yin, Y. (2000). Mining frequent patterns without candidate generation. In *Proceedings of the 2000 ACM SIGMOD international conference on Management of data* (pp. 1-12).

Hassan, M. K., Aliyu, S., Huda, M., & Rashid, M. (2019). A Survey on Islamic Finance and Acconting Standarts. *Borsa İstanbul Review, 19*(1), 1–13. DOI: 10.1016/j.bir.2019.07.006

Kshatri, S. S., Singh, D., Goswami, T., & Sinha, G. R. (2023). *Introduction to statistical modeling in machine learning: A case study. Statistical Modeling in Machine Learning*. Elsevier. DOI: 10.1016/B978-0-323-91776-6.00007-5

Nanopoulos, A., Papadopoulos, A. N., & Manolopoulos, Y. (2007). Mining association rules in very large clustered domains. *Information Systems, 32*(6), 649–669. DOI: 10.1016/j.is.2006.04.002

Özsoy, Ş. (2012). *Sağlam bankacılık modeli ile katılım bankacılığına giriş*. Kuveyt Türk Yayınları.

Pireh, M. (2008). *Definition of Islamic Financial Instruments, Islamic Finance Expert Securites & Exchange Organization (SEO), 21*. Tahran.

Raza, S., & Ding, C. (2021). News recommender system: A review of recent progress, challenges, and opportunities. *Artificial Intelligence Review, 55*(1), 749–800. DOI: 10.1007/s10462-021-10043-x PMID: 34305252

Roy, D., & Dutta, M. (2022). A systematic review and research perspective on recommender systems. *Journal of Big Data, 9*(1), 59. DOI: 10.1186/s40537-022-00592-5

Sarath, K. N. V. D., & Ravi, V. (2013). Association rule mining using binary particle swarm optimization. *Engineering Applications of Artificial Intelligence, 26*(8), 1832–1840. DOI: 10.1016/j.engappai.2013.06.003

Sarath, K. N. V. D., & Ravi, V. (2013). Association rule mining using binary particle swarm optimization. *Engineering Applications of Artificial Intelligence, 26*(8), 1832–1840. DOI: 10.1016/j.engappai.2013.06.003

Shafii, Z., Zakaria, N., Sairally, B. S., Shaharuddin, A., Hussain, L., & Zuki, M. S. M. (2013). *An Appraisal of the Principles Underlying International Financial Reporting Standards (IFRS): A Shari'ah Perspective-Part 1*. International Shari'ah Research Academy for Islamic Finance (ISRA).

Shalev-Shwartz, S., & Ben-David, S. (2014). *Understanding Machine Learning: From Theory to Algorithms*. Cambridge University Press. DOI: 10.1017/CBO9781107298019

Srivastava, T., Mullick, I., & Bedi, J. (2024). Association mining based deep learning approach for financial time-series forecasting. *Applied Soft Computing, 155*, 111469. DOI: 10.1016/j.asoc.2024.111469

Ünlü, R. (2022). An Assessment of imbalanced control chart pattern recognition by artificial neural networks. İçinde Research Anthology on Artificial Neural Network Applications (ss. 683-702). IGI Global. https://www.igi-global.com/chapter/an-assessment-of-imbalanced-control-chart-pattern-recognition-by-artificial-neural-networks/288982

Ünlü, R., & Xanthopoulos, P. (2021). A reduced variance unsupervised ensemble learning algorithm based on modern portfolio theory. *Expert Systems with Applications, 180*, 115085. DOI: 10.1016/j.eswa.2021.115085

Witten, I. H., Frank, E., Hall, M. A., & Pal, C. J. (2017). *Data Mining: Practical Machine Learning Tools and Techniques*. Morgan Kaufmann.

Zhou, K., Luo, L., & Chen, T. (2023). The Influence of Online Media on College Students' Self-identity in Mobile Learning Environment. In 2023 4th International Conference on Education, Knowledge and Information Management (ICEKIM 2023), 1195-1203.

Zhou, L., & Yau, S. (2007). Efficient association rule mining among both frequent and infrequent items. *Computers & Mathematics with Applications (Oxford, England)*, *54*(5), 737–749. DOI: 10.1016/j.camwa.2007.02.010

Chapter 2
Investigation of Artificial Intelligence Models (AI) in Shariah Auditing of Islamic and Conventional Financing

Mohamed Sadok Gassouma
https://orcid.org/0000-0002-0932-9326
Ez-Zitouna University, Tunisia

ABSTRACT

This paper aims to conduct a Shariah classification of various financing transactions granted to Tunisian commercial institutions listed on the stock exchange by Tunisian banks, whether Islamic or conventional. This classification consists of a Shariah automated audit issued from the artificial intelligence based on a set of Islamic financial jurisprudential principles to verify compliance in banking transactions. The study maintains complete objectivity and neutrality toward all banks, regardless of their affiliations and orientations The results indicate that the Logistic Regression Model proved to be the most effective, as it generated the lowest margin of error. However, the Islamic legitimacy of financing primarily depends on factors such as productive work, the principle of profit loss sharing, balance between money and goods, and the avoidance of debt-based trading. These artificial intelligence models can be programmed into computerized systems, making

DOI: 10.4018/979-8-3693-8079-6.ch002

them automated and capable of real-time auditing

1. INTRODUCTION

Given the differences among members of society regarding the legitimacy of Islamic financing compared to conventional financing, and considering the similarities between both as shown by Gassouma et al (2022), contemporary financial pioneers have turned to Sharia compliance, also known as Sharia auditing of Islamic financial transactions.

Sharia auditing is often directed solely at declared Islamic financing. From the perspective of Sharia auditors, conventional finance is entirely illegitimate and, therefore, not subject to Sharia auditing, effectively placing it outside the scope of Islamic law. However, if conventional finance is inherently illegitimate, why has the Islamic world adopted it as a fundamental reference, adapted it, and framed it within a Sharia context? The response from Islamic scholars and banking experts is as follows: Islamic finance is essentially conventional finance that has been adapted to comply with Sharia, evolved over time, and no longer necessitates reverting to conventional methods. In fact, they argue that conventional finance is no longer viable in the Islamic world. (Sidiqqi, 1982).

Our discussion in this paper falls within these considerations. We acknowledge that the current state of Islamic finance requires further auditing and development. Additionally, Islamic finance, as practiced in contemporary banks, may deviate from true Islamic financial principles, often resembling financial transactions in form but not in substance. Conversely, conventional finance can, in essence, align with Islamic financial principles while differing in form. (Askari and Rehman, 2010).

By "form," we refer to the structuring and binding nature of contracts for all parties involved. Islamic financial transactions and financing follow a legally structured framework of guidelines that ensure compliance with Sharia principles. However, in many cases, Islamic financing may adopt a rigorous legal and regulatory framework while lacking substantive Sharia compliance.

The content of financial transactions involves how financial mechanisms are structured, including profit calculation methods, investment rates, deferred payment pricing, debt repayment amounts, profit margins, and other financial factors. Sometimes, conventional financial mechanisms are merely rebranded with Islamic terminology while maintaining their original conventional substance.

On the other hand, conventional finance may sometimes contain genuinely Sharia-compliant elements, particularly in profit and interest calculations. Many conventional banks offer loans with a profit margin that aligns with asset returns rather than monetary returns. However, the terminology used is not considered Sharia-compliant, as such profits are referred to as "interest," and the contract itself is structured as a loan agreement rather than a sale or partnership contract. (Gassouma, 2023.a)

Thus, what differentiates a loan contract from a sale, purchase, or Mudaraba (profit-sharing) contract? Does every loan in conventional finance constitute the usurious lending (riba) that is prohibited by God in "Surah Al-An'am" of the Quran? Are all conventional financial transactions non-compliant with Sharia, or do some align with its principles? Furthermore, is Islamic finance inherently Sharia-compliant, or does it also require Sharia auditing?

Many researchers argue that auditing conventional finance is unnecessary; instead, the focus should be on auditing Islamic finance to ensure its superiority over conventional finance. This perspective implies that the Islamic world needs more division and rigidity to uphold its religious values. (Arif, 1985) However, shouldn't the true goal of upholding Islam involve unity and the development of new financial models that accommodate both Muslim and non-Muslim societies?

The objective of this paper is to establish a unified Sharia-compliant financial model that integrates both Islamic and conventional finance, creating a financial system that adheres to Sharia principles within a modern civil framework.

To achieve this, we propose extending Sharia auditing to both Islamic and conventional finance. However, current Sharia auditing methods sometimes succeed and sometimes fail due to their subjective nature. Therefore, in this paper, we adopt a standardized auditing approach based on Sharia principles rather than an inductive approach relying on surveys and empirical analysis. Our methodology is analytical and objective: for each essential Sharia principle in financial transactions, we provide a theoretical framework, assess it using standardized criteria, and apply measurable benchmarks.

This theoretical and quantitative assessment is not sufficient for making immediate, automatic decisions. Thus, we introduce standardized models that integrate all relevant Sharia principles, allowing us to objectively determine the legitimacy of financial transactions. These models enhance efficiency and improve the accuracy of results.

These models fall under the category of artificial intelligence (AI) models (Han and Kamber, 2006). They consist of statistical rules that can later be programmed into information systems, making them applicable in both Islamic and conventional banks for Sharia auditing.

Additionally, these models can be utilized by financial institutions that have benefited from either Islamic or conventional financing to assess their compliance with Sharia. Furthermore, Islamic banks do not derive the legitimacy of their financial transactions solely from internal policies but rather from the institutions they interact with. This is because the essence of an Islamic financial transaction lies not in its structure but in its impact on external financial markets and the surrounding environment.

For example, the prohibition of riba (usury) can only be truly assessed by auditing the client. So, only the client's financial behaviour that ultimately determines whether the bank's financing is conformed or no to shariah principles.

An Islamic bank's responsibility does not end with disbursing financing; rather, its obligations begin at that moment. An Islamic bank has a collective responsibility toward the external world, ensuring that its transactions derive legitimacy not only from internal compliance but also from the financial conduct of the beneficiaries throughout the duration of the contract.

This paper follows a two-step approach. First, we conduct an inductive and quantitative analysis of key Sharia principles, including risk-sharing (al-ghunm bil-ghurm), the importance of labor, prohibition of hoarding and encouragement of spending, prohibition of monopolies and emphasis on justice, and the prohibition of riba (usury). Each principle is examined from an economic, financial, and jurisprudential perspective, concluding with a measurable formula for verification.

In the second part, we apply these models, conduct quantitative analysis, and interpret the findings, ultimately concluding with recommendations and final observations.

2. STANDARDIZED THEORIZATION OF SHARIA PRINCIPLES

2.1 Labor

Labor is one of the fundamental pillars of Islam, as humanity cannot achieve stewardship and development without work. (Adesina, 2020).

In economic terms, labor is primarily based on trade, as highlighted by Allah in the Quran:

"O you who have believed, shall I guide you to a transaction that will save you from a painful punishment " (Surah Ash-Shura, Ayah 26). This verse links faith with commerce, reinforcing the importance of labor in earning a livelihood. To honor work, the Prophet Muhammad (peace be upon him) have compared it to jihad (struggle) in the way of Allah, recognizing that labor entails both physical and mental effort. Labor is the key driver of economic growth and a fundamental factor in economic and monetary balance. The Prophet (peace be upon him) said:

"The worker who strives truthfully for his sustenance is like a warrior in the path of Allah until he returns home."

Labor is thus a fundamental Sharia objective for achieving economic growth through production. All production eventually turns into either final consumption, intermediate consumption, investment, or savings. Labor serves as the primary engine of wealth creation. When money is integrated with labor, it generates added value in the form of goods or services. Hence, labor plays a role in preventing capital erosion over time. (Asare et al, 2017)

Islamic banks must ensure that the financing they provide is based on labor and its value. Unlike conventional banks, Islamic banks calculate their profits based on the labor value in every financed asset, considering that labor is assessed through the added value generated by investment or consumption.

2.2 The Principle of Money-Goods Equilibrium (Al-Ghunm bil Ghurm)

The principle of *Al-Ghunm bil Ghurm* (money-goods equilibrium) reflects the economic equilibrium between the supply of idle money and the demand for real assets. In an economic financial system, every unit of money supplied must correspond to a demand for goods, services, or other forms of capital. In contrast, the conventional financial system allows money to be exchanged not only for assets but also for other forms of money. (Khan and Mirakhor, 1989)

John Maynard Keynes theorized the supply and demand for money. According to him, the money supply consists of all circulating funds held in banks, either in savings accounts (investment accounts) or current accounts. The demand for money, on the other hand, depends on its intended use and is divided into three categories:

1. **Transactional money:** Used for investment and fixed consumption.
2. **Precautionary money:** Held for emergency situations.
3. **Speculative money:** Used for profit-seeking in financial markets.

Keynesian theory (1930), suggests that the equilibrium price in the goods and services market depends on the balance between supply (production) and demand (consumption). Hicks (1937) later refined this model into the IS-LM framework, which captures the interaction between the goods and money markets.

In contrast, contemporary Islamic economists such as Askari and Mirakhor (2017) have developed Islamic economic models that mirror Keynesian and IS-LM theories but with a fundamental difference: they exclude speculation on money. These Islamic models aim to curb financial speculation and the creation of artificial wealth.

2.3 The Prohibition of Hoarding and the Encouragement of Spending

Spending is one of the key drivers of investment, as it opposes hoarding, which is prohibited in Islam. Hoarded money refers to wealth that is saved but does not contribute to economic activity. Any wealth that exceeds annual ordinary expenses, after deducting zakat and taxes, is considered idle money and must be reinjected into the economy.

Allah warns against hoarding in the Quran:

"And those who hoard gold and silver and do not spend it in the way of Allah—give them tidings of a painful punishment. On the Day it will be heated in the fire of Hell and seared onto their foreheads, their sides, and

their backs: 'This is what you hoarded for yourselves, so taste what you used to hoard.'" (Surah At-Tawbah, Ayahs 34-35).

Ibn Khaldun (1377) also highlighted that depriving individuals of their basic needs can lead to negative economic consequences. When people are forced to suppress their desires, they tend to hoard wealth instead of spending it, which in turn hampers economic growth. This can lead to an increase in money circulation without corresponding labor and production, potentially resulting in the rise of usurious transactions.

Keynes' 1930 theory aligns with Islamic financial objectives, as he argued that saving does not necessarily lead to investment; rather, it can divert funds from consumption, negatively impacting national economies. According to Keynes, consumption and investment are complementary, not substitutes. Increased consumption encourages investors to create new projects to meet demand, fostering economic expansion. This idea is similar to the Islamic principles of consumer and investment spending.

Islamic banks and financial markets should therefore design financial products that align with this principle. Before issuing financing, an Islamic bank should not solely focus on ensuring its transactions are free from usury (*riba*), but also assess whether the funded activities contribute to sustainability, wealth circulation, and social benefits.

2.4 The Prohibition of Fraud and Deception

Fraud and deception are major causes of social injustice. These unethical practices manifest in various forms, particularly in fraudulent pricing and dishonest trade practices. The Quran explicitly forbids fraud:

"And to Madyan (the people of Midian), We sent their brother Shu'ayb. He said, 'O my people, worship Allah; you have no deity other than Him. Do not decrease the measure and weight in trade. Indeed, I see you in prosperity, but I fear for you the punishment of an all-encompassing day. O my people, give full measure and weight in justice and do not deprive people of their due, and do not commit abuse on the earth, spreading corruption.'" (Surah Hud, Ayahs 84-85).

Islam commands fair measurement and forbids any reduction that distorts economic balance. Fraudulent practices can disrupt supply and demand dynamics, leading to artificially inflated prices.

This principle extends to accounting irregularities and financial misrepresentation. Just as the balance between weight and price must be maintained, financial reporting should reflect the actual balance between production and monetary supply. Economic imbalances resulting from financial manipulations can lead to artificial asset inflation (Benhamed and Gassouma, 2023) and unearned wealth accumulation, exacerbating economic inequality. (Chapara, 1989).

2.5 The Prohibition of Usury (Riba)

The essence of *riba* (usury) lies in increasing wealth solely through time, without engaging in productive economic activity. This results in a surplus of money created for the purpose of trading time, under the pretext that money loses value over time.

In reality, money itself does not carry an inherent cost; its cost can only be determined through its intended use. If money is used for investment, its cost corresponds to the investment expenses. If it is used for consumption, its cost is the cost of the consumed goods or services and this is over a given period. Thus, the cost of money is determined by the growth of its utilization over time, which varies across different asset classes. (Mashhour, 1991).

The most prevalent form of *riba* today is *riba al-nasi'ah*, known in pre-Islamic Arabia as "deferred usury." In this system, a debtor was required to either repay a loan or extend it with additional interest. While modern conventional banks justify interest as compensation for the declining value of money over time, Islamic finance maintains that money should not generate returns without economic activity.

This raises the question: What is the difference between *riba* and interest? Is *riba* equivalent to interest, or is it merely a subset of it?

Interest represents any increase in financial or real assets. If the increase stems from asset production, it is considered commercial profit (*profit margin*), permissible in Islamic finance. However, if the increase arises solely over the time without productive economic activity, it is classified as *riba*.

This prompts the fundamental inquiry: Do conventional and Islamic banks engage in *riba* or legitimate profit-making when they issue loans for asset financing?

3. APPLICATION OF ARTIFICIAL INTELLIGENCE MODELS

3.1 Sample

The study sample consists of a set of Tunisian institutions that have benefited from bank financing, divided into institutions that received Islamic financing and others that obtained conventional financing. The period covered extends from the year 2000 to 2022. The data is sourced from financial statements and annexes available in the Tunis Stock Exchange database.

3.2 Variables

Based on the theoretical framework, we extracted seven factors representing compliance with Shariah principles. Each Shariah principle is assigned a benchmark or threshold, which is then transformed into a variable.

The First Principle is the money-goods equilibrium Principle (Al-Ghunm bil Ghurm): This principle ensures a balance between the real economy and the monetary economy, preventing the creation of virtual usurious money and financial inflation. We propose this ratio to verify the equilibrium between goods measured by Fixed assets and money indicated by long term funds: (Gassouma, 2022.a)

$$\textit{Money} - \textit{Goods equilibium} = \left(\frac{\textbf{Fixed Assets}}{\textbf{Equity} + \textbf{Non} - \textbf{Current Liabilities}}\right) - 1$$

The higher the value of this variable, the more the principle is violated, leading to a disconnect between the real and monetary economies.

The Second Principle is The Labor: Labor is the sole creator of real wealth and enables the bank to generate non-usurious profits, making its transactions fully Shariah-compliant. According to Adesina (2020), the labor factor is expressed as:

$$\textbf{LABOR} = \frac{\textbf{Revenues} - (\textbf{Operating Expenses} - \textbf{Employment Costs})}{\textbf{Employment Costs}}$$

The higher this ratio, the greater the value of labor, ensuring a more compliant transaction.

The Third Principle is Prohibition of Hoarding and Encouragement of Spending: Avoiding hoarding fosters social benefit and promotes investment through consumption and exploitation. If the financier hoards wealth, their returns weaken, affecting the borrower's repayment capacity. We suppose that when the net cash exceed the third (1/3) of the total assets, the enterprise fall into the Hoarding money. We propose as hoarding factor: Gassouma (2023)

$$\textbf{Hoarding money} = \left(\frac{\textbf{Net cash}}{\textbf{Total Assets}}\right) - 0.33$$

The threshold of one-third is set as a maximum liquidity criterion based on Shariah principles.

The Fourth Principle: Prohibition of Merchandise speculation: Speculation is forbidden as it disrupts liquidity and affects the repayment cycle of borrowers. We consider that when the merchandise detention exceeds the third of total assets, the enterprise practice speculation. The factor is defined as: (Gassouma, 2022)

$$\textbf{Speculation} = \left(\frac{\textbf{Inventory}}{\textbf{Total Assets}}\right) - 0.33$$

Fifth Principle: Prohibition of Debt Trading and Purchases: The more deferred sales a financed institution engages in, the more its resources become questionable, involving debt trading. The debt purchase variable is calculated as:

$$\textbf{Debt Purshase} = \frac{\textbf{Trade Receivable}}{\textbf{Total Assets}} - 0.33$$

The Sixth Principle is the Prevention of Fraud: The Fraud leads to economic losses, particularly for the lending bank. To measure fraud, we adopted the Kothari et al. (2005) model, which calculates fraud value and earnings manipulation. The model is given as:

$$\frac{ACT_{i,t}}{TA_{i,t-1}} = \alpha 0 \times \frac{1}{TA_{i,t}} + \alpha 1 \times \frac{FA_{i,t}}{TA_{i,t}}$$

$$+ \alpha 2 \times \frac{\left(\Delta Turnover_{i,t} - \Delta CUD_{i,t}\right)}{TA_{i,t}} + \alpha 3 \times \frac{NI_{i,t-1}}{TA_{i,t-1}}$$

Where:

- **ACT** = Total Fraud Value = Net Income - Cash Flows
- **TA** = Total Assets
- **FA** = Fixed Assets
- **Turnover** = Revenues
- **CUD** = Receivables
- **NI** = Net Income

After performing linear regression and computing the coefficients, we extracted the error margin for each financing. The error margin in the results indicates the extent of fraud.

The Seventh Principle: Prohibition of Usury (Riba): To avoid usury, banks assign a profit margin to their financing, reflecting the return on financed assets rather than the return on money itself. We first calculate the applied financing rate as: (Gassouma and Ghroubi, 2021; Gassouma, 2023)

$$\textbf{Applied Interest Rate} = \frac{\textbf{Intesrest Paid}}{\textbf{Debt}}$$

Then, we estimate the theoretical profit margin based on the return on debt-financed assets:

$$\textbf{Theorical Intesrest rate} = \left(\frac{\textbf{Accounting Profit}}{\textbf{Total Fixed Assets}}\right) \times \left(\frac{\textbf{Debt Value}}{\textbf{Total Assets}}\right)$$

The further the applied financing rate deviates from the theoretical profit margin, the more the bank relies on usurious interest rather than asset-based returns.

Table 1. Descriptive statistics

Variable	N	Minimum	Maximum	Mean	Standard Deviation
Money-Goods equilibrium	2150	0.00	4.01	0.7477	0.82367
Debt Trading	2150	-0.032	0.23	-0.1145	0.17504
Speculation	2150	-0.33	0.51	-0.0449	0.22751
Hoarding	2150	-0.077	-0.29	-0.4841	0.14356
Usury (Riba)	2150	-0.04	0.21	0.0517	0.05803
Fraud	2150	0.00	0.21	0.0799	0.05944

3.3 Design of Artificial Intelligence Models

3.3.1 Discriminant Analysis Model

This model aims to distinguish between Islamic and conventional financing based on a set of weighted factors. Weighting allows us to assign importance to each factor in differentiation. The model also enables us to extract the probability of financing being either Islamic or conventional. (Atlman, 1968)

Upon applying this model, it yielded a positive discrimination rate of 63%, which is considered a good result. This means that 33% of the financings were misclassified, with 35.7% of conventional financings classified as Islamic and 38.5% of Islamic financings classified as conventional.

Table 2. Classification results of discriminant analysis

Declared Financing	Predicted Financing	
	Conventional	Islamic
Conventional	64.3%	35.7%
Islamic	38.5%	61.5%
Overall Accuracy	**63.0%**	

This indicates significant overlap between Islamic and conventional financing. As a result, we decided to develop an equation to determine the influential factors that contribute to fully compliant Islamic financing:

Table 3. Impact principles on legitimacy

Factor	Islamic Legitimacy
Labor	0.424
Money-Goods equilibrium	-0.433
Debt Trading	-0.081
Speculation	-0.570
Hoarding	-0.304
Usury (Riba)	-0.139
Fraud	-0.059

We observe that the most important factor in Islamic legitimacy of transaction is the speculation of merchandise prohibition, then is the money-Goods equilibrium and Labor. So, This model helps identify the key elements affecting the legitimacy of financing according to Islamic finance principles.

Figure 1. Discriminant probability of Islamic financing

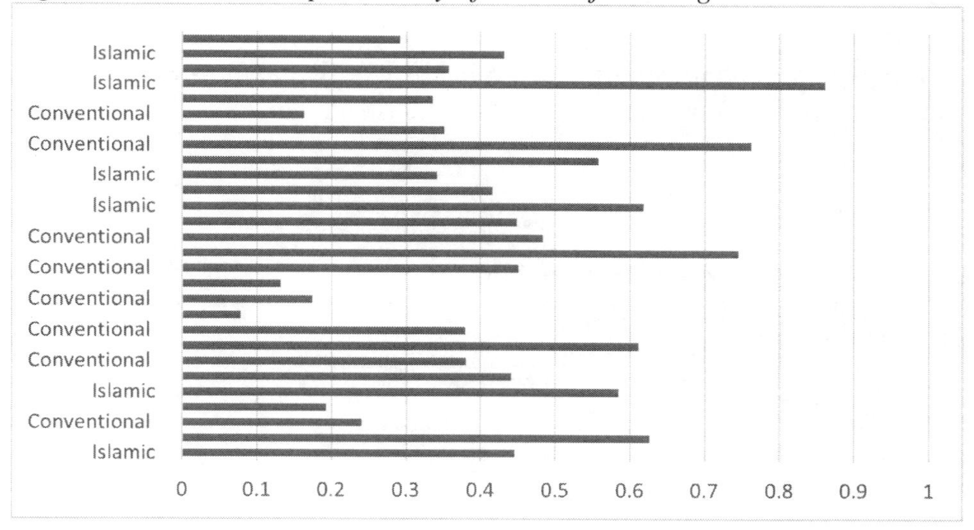

By computing the Islamic probability of financing, we observe that there are some conventional financings having more Islamic probability that the Islamic ones. This leads to conclude that Islamic legitimacy of transaction is not linked to the bank origine (Islamic or conventional) but to the nature of transaction and the principles conforming.

3.3.2 Logistic Regression Model

This model aims to calculate the probability that both types of financing are Islamic. It also helps distinguish between Islamic and conventional financing through different weightings for each Sharia-compliant factor expressed as a variable. (Rajhi and Gassouma, 2007)

The model showed that the correct classification rate is approximately 65%, with 33.1% of conventional financing classified as Islamic and 50% of Islamic financing classified as conventional. This indicates a significant overlap between both types of financing.

Table 4. Classification result of logistic regression

Declared Financing	Conventional	Islamic
Conventional	76.9%	33.1%
Islamic	50%	50%
Correct Classification	**65%**	

Below is a graph illustrating the probability of financing being classified as Islamic for both conventional and Islamic financing:

Figure 2. Logistic probability of Islamic financing

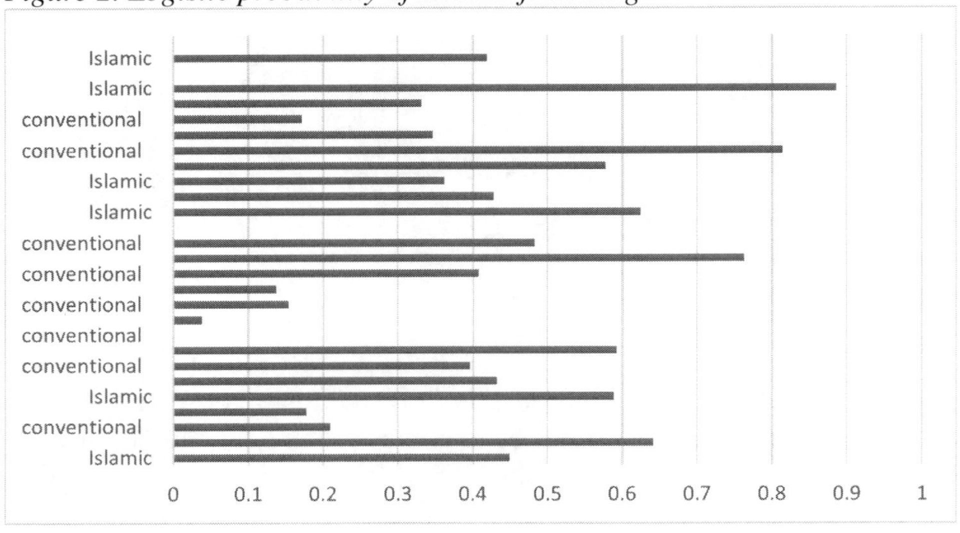

Similarly, as in the discriminant analysis, the logistic regression show that there are some conventional financings having an important Islamic probability than Islamic ones and reciprocally.

The role of each variable in contributing to the legitimacy of financing is summarized in the following table:

Table 5. Discriminant power of Sharia principles of logistic regression

Sharia Principles	Impact on Legitimacy	Weighting
Labor	-0.043	0.958
Money-Goods disequilibrium	-0.594	1.811
Debt Trading	-0.316	0.729
Speculation	-2.568	0.077
Hoarding	-2.204	0.110
Usury (Riba)	-1.526	0.217
Fraud	-0.974	0.378

The table.5 show that the most factor contributing in Islamic legitimacy is Money-goods equilibrium then is "Labor", and in the third rank is debt trading.

3.3.3 K-Neighbors-Means Clustering Model

This model differs from the previous ones as it is non-weighted and does not assign different effects to Sharia principles and variables in determining the probability of legitimacy. Instead, it combines all variables under a single category financing legitimacy and calculates the distance between each principle and its legitimacy by averaging the values. Then, it classifies each financing type based on its proximity to legitimacy. (Atlman, 1992)

The results of this model summarize the different probabilities of Islamic classification for various financing cases as follows:

Figure 3. KNN probability of Islamic funding

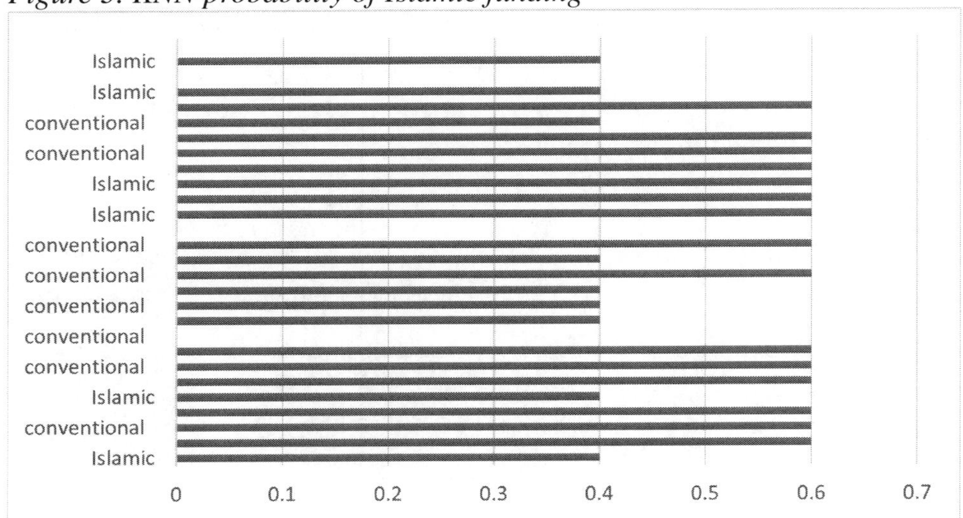

3.3.4 The Convergence Curve Between Models: ROC Curve

This curve aims to highlight the most effective model. The higher the curve of a model, the more effective it is. Below is the convergence curve: (Hand et al, 2003)

Figure 4. ROC curve

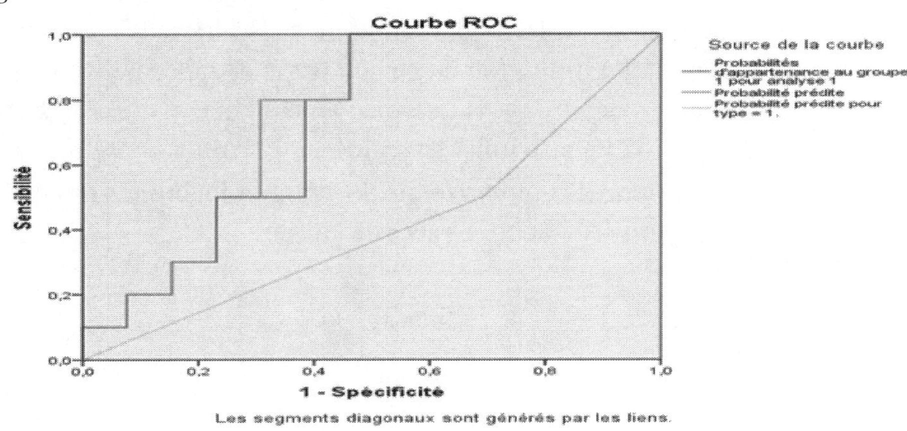

Based on this model, we found that the most effective model is the logistic regression model. Therefore, we will select this model for further analysis and results interpretation, relying on the findings presented above. So, we test now the contribution of each factor in Islamic Legitimacy of transaction. (Table 5)

4. MAIN RESULTS

we conclude that all jurisprudential Islamic factors and principals have a negative effect on the Islamic legitimacy of financing, except for the labor control, having a positive impact. These results are highly reasonable since the increase of all prohibiting factors failing reduces the legitimacy of financing, whereas adhering to them enhances legitimacy.

The most factor contributing to the Islamic legitimacy the principle of "Money-Goods disequilibrium", leading to maintains balance between the real economy and the monetary economy. The effect of this last factor on financing legitimacy is negative, meaning that as fixed assets move further away from self-financing and non-current liabilities, the real economy diverges from the monetary economy, reducing legitimacy.

This guarantees that financing is directed towards tangible assets and not merely bound by contractual terms, as contracts may state that financing is asset-based while the reality differs. This factor is essential for eliminating ambiguity in transactions.

In the second rank, we find the labor control, which have a positive effect. The more labor increases, the more financing, whether conventional or Islamic, becomes genuinely Islamic. Labor is one of the most critical pillars of Islamic finance because it creates added value to assets, increasing their worth and profitability. The more labor input increases, the more real profits are achieved, contributing to making bank financing productive rather than interest-based.

In the third degree, we find the debt trading that has a negatively effects on financing legitimacy, whereas limiting debt trading serves to enhance it .It is clear that institutions benefiting from banking financing based on trading customers' debts, inflates liquidity and capital while generating unproductive profits, thereby reducing the legitimacy of the institution and its financing.

Fraud is the fourth factor contributing to Islamic legitimacy. It also has a negative impact on financing legitimacy. When the financing beneficiary engages in fraudulent activities, financing automatically loses its legitimacy because the profits become manipulated and inconsistent with reality, causing prices to deviate from their theoretical levels and disrupting the balance between the real and monetary economy. To ensure financing legitimacy, banks must verify that beneficiaries are free from fraudulent practices, guaranteeing that their profits reflect actual economic realities.

Then, money speculation and hoarding have an overwhelmingly negative impact on financing legitimacy. The more a financing beneficiary hoards wealth, the less productive and beneficial the financing becomes, failing to generate real profits and instead fostering usury. Similarly, monopolization contributes to illegitimacy for the same reasons.

As result, the no-respect of these factors lead to money usury that limits the legitimacy. The prohibition of usury (riba) is one of the fundamental pillars of financing legitimacy, observed in the last rank. However, riba is a consequence of failing to adhere to other Shariah controls. When labor is absent and the hoarding increases, production decreases, and real profit declines. As a result, the borrowers, when he repays his debt with interest/ profit, this last systematically don't be linked to the real assets and so, it will be qualified as an interest and not profit, even if the bank is an Islamic bank.

As mentioned earlier, riba is the difference between the profit rate of capital invested in tangible assets (i.e., the production return from financing assets) and the rental rate applied by the bank. If the latter converges with the theoretical profit rate, it represents an Islamic profit rate; otherwise, it constitutes usurious interest. The variable of **riba** represents the deviation of the bank's rental rate from the Islamic rate. We observe that the effect of riba is negative, meaning that as riba increases, financing legitimacy decreases, making it illegitimate.

The study also examined the probability of Islamic financing legitimacy through probability charts, demonstrating that both conventional and Islamic financing can achieve legitimacy to varying degrees, with Islamic financing having a higher probability of legitimacy. Both types of financing must adhere to the outlined jurisprudential controls to be considered Shariah-compliant.

5. CONCLUSION

This study introduced an automated classification system for both Islamic and conventional financing as they currently exist. The objective was to establish AI-based standards to facilitate Shariah auditing, making it an objective and systematic process. This method helps Shariah auditors and regulatory bodies, whether in banks or financed institutions, to avoid errors while saving time and costs.

The most effective and accurate model in this field was found to be logistic regression, as it provided the best classification and auditing results. The study demonstrated that both Islamic and conventional financing can align with Shariah principles but may also differ. Both financing types possess inherent legitimacy depending on transactions, with probability variations sometimes favouring conventional financing over Islamic financing. The legitimacy of financial transactions does not depend on contract formats, a bank's trade name, or its ideological orientation but rather on compliance with a set of financial Shariah controls.

The study further highlighted that financing legitimacy primarily stems from the principle of money-Goods equilibrium "Al-Ghunm bil-Ghurm", the balance between the monetary and real economy, and the significance of labor in generating added value. Respecting these controls ensures financing legitimacy, which in turn ensures the legitimacy of the bank.

Thus, for an Islamic or conventional bank to demonstrate its legitimacy and be considered genuinely Islamic, it must engage in social responsibility toward its clients. The responsibility of an Islamic bank is not limited to internal management and compliance with Sharia contracts. Compliance is also understood through the financing process and the behaviour of the client. The Shariah compliance of financing begins with the request for financing and continues until the partnership or debt between the bank and the client ends."

Ultimately, we affirm that a bank's legitimacy is derived from the transactions of its financing beneficiaries rather than from the bank itself, as transactions are not just contractual events but have economic repercussions that define their legitimacy.

REFERENCES

Adesina, K. S. (2020). How diversification affects bank performance: The role of human capital. *Economic Modelling*.

Ahmad, Z. (1984). Concept and Models of Islamic Banking: An Assessment," paper presented at the Seminar on Islamization of Banking, Karachi.

Altman, E. (1968). Financial ratios, discriminant analysis and the prediction of corporate bankruptcy. Journal of financial, pp 189-209. DOI: 10.1111/j.1540-6261.1968.tb00843.x

Altman, N. S. (1992). An introduction to kernel and nearest-neighbor non-parametric regression. *The American Statistician*, *46*(3), 175–185. DOI: 10.1080/00031305.1992.10475879

Arif, M. (1985). Toward the sharia paradigm of Islamic Economics, the beginning of a scientific revolution, *The American of Islamic Social Science*-vol 2.

Asare, N., Alhassan, A. L., Asamoah, M. E., & Ntow-Gyamfi, M. (2017). Intellectual capital and profitability in an emerging insurance market. *Journal of Economic and Administrative Sciences*, *33*(1), 2–9. DOI: 10.1108/JEAS-06-2016-0016

Askari, H., & Mirakhor, A. (2017). *Ideal Islamic Economy*. Book.

Askari, H., & Rehman, S. (2010). *How Islamic are Islamic Countries?* (Vol. 10). Global Economy Journal.

Benhamed, A., & Gassouma, M. S. (2023). Preventing Oil Shock Inflation: Sustainable Development Mechanisms vs. Islamic Mechanisms. *Sustainability (Basel)*, *15*(12), 2. DOI: 10.3390/su15129837

Chapra, M. U. (1982). Money and Banking in an Islamic Economy. In Ariff, M. (Ed.), *Monetary and Fiscal Economics of Islam*. International Centre for Research in Islamic Economics.

Gassouma, M. S. (2023). Applications of an Alternative Islamic Profit Rate to Interest Rates: Examining Its Impact on Investment and Inflation, *2nd International Conference on Islamic Finance*, Tunisia – Hammamet, University of Zitouna.

Gassouma, M. S. (2023a). Empirical Modeling of Sharia Compliance in Islamic Financial Transactions: A Case Study on the Alignment of Islamic and Conventional Finance with Sharia Standards, 6th International Scientific Conference Islam and Contemporary Issues, Turkey – Antalya. Euro-Arab Organization for Water and Desert & University of Zitouna.

Gassouma, M.S., Benahmed, A., Montasser, G. (2022). Investigating similarities between Islamic and conventional banks in GCC countries: a dynamic time warping approach. *International Journal of Islamic and Middle Eastern Finance and Management.*

Gassouma, M.S., Ghroubi, M. (2021). Discriminating between Islamic and Conventional banks in term of cost efficiency with combination of credit risk and interest rate margin in the GCC countries: Does Arab revolution matter? *ACRN Oxford Journal of Finance and Risk Perspectives, 10.*

Gassouma, M.S and Rajhi, M.T. (2007). Evaluation de Data Mining pour la classification des entreprises industrielles Tunisiennes dans le cadre du credit scoring. Revue de l'association francophone de management électronique.

Han, J., & Kamber, M. (2006). *Data Mining: Concepts and Techniques* (2nd ed.). Morgan Kaufmann.

Hand, D. J., Till, R. J., & Patil, S. (2001). A simple generalization of the area under the ROC curve for multiple class classification problems. *Machine Learning*, *45*(2), 171–186. DOI: 10.1023/A:1010920819831

Hicks, J. R. (1937). Mr. Keynes and the 'Classics': A Suggested Interpretation. *Econometrica*, *5*(2), 147–159. DOI: 10.2307/1907242

Ibn khaldoun (1377). El Mukaddimah. Book.

Ibrahim, A. S., Zaki, A., & Al-R., A. (2023). Applied Ijtihad of Sharia Rulings: Contemporary Financial Transactions as a Model. *IUG Journal of Sharia & Law Studies, 31*(1).

Keynes, J. M. (1936). *The General Theory of Employment, Interest and Money*. Book.

Khan, M. S. (1986). Islamic Interest-Free Banking: A Theoretical Analysis. *Staff Papers - International Monetary Fund. International Monetary Fund, 33*(1), 1–27. DOI: 10.2307/3866920

Khan, M. S., & Mirakhor, M. (1989). The financial system of monetary policy in an Islamic Economy. *Islamic Economic, Vol, 1*(1), 39–57. DOI: 10.4197/islec.1-1.2

Siddiqi, M. (1982). *Nejatullah.* Islamic Approaches to Money, Banking and Monetary Policy.

iddīqī, M. N. (1981). Muslim economic thinking: A survey of contemporary literature. Islamic economics series.

Chapter 3
Geoeconomic and Geopolitical Transformation Projects:
The Islamic Finance and Related Technology Strategic Vision in the Lebanese Context (IFSVLC)

Antoine Trad
https://orcid.org/0000-0002-4199-6970
Independent Researcher, France

ABSTRACT

In this chapter related to Geoeconomical and Geopolitical (GG), the author uses the Applied Holistic Mathematical Model (AHMM) for GG(AHMM4GG) that can be applied for economy, finance, societal, and business transformations, assessment of GG Risks (GGRisk), GG Problems (GGProblem) and possible eXtremely High Failure Rates (XHFR). These GGRisks can be mitigated by the use of Critical Success Factors (CSFs) (Trad, & Kalpić, 2020a), which can be used in the context of conflictual regions evolution, like the Middle East Area (MEA) and more specifically Lebanon. This chapter is based on a unique mixed research method that is supported by a mainly qualitative research module (Gunasekare, 2015). The AHMM4GG for conflictual and GG Transformation Projects (simply Projects) uses a scripting environment that can be adopted by a Project and for that goal the author

DOI: 10.4018/979-8-3693-8079-6.ch003

uses it to research the The Islamic Finance and Related Technology Strategic Vision in the Lebanese Context (IFSVLC)

INTRODUCTION

In this chapter related to Geoeconomical and Geopolitical (GG), the author uses the Applied Holistic Mathematical Model (AHMM) for GG(AHMM4GG) that can be applied for economy, finance, societal, and business transformations, assessment of GG Risks (GGRisk), GG Problems (GGProblem) and possible eXtremely High Failure Rates (XHFR). These GGRisks can be mitigated by the use of Critical Success Factors (CSFs) (Trad, & Kalpić, 2020a), which can be used in the context of conflictual regions evolution, like the Middle East Area (MEA) and more specifically Lebanon. This chapter is based on a unique mixed research method that is supported by a mainly qualitative research module (Gunasekare, 2015). The AHMM4GG for conflictual and GG Transformation Projects (simply Projects) uses a scripting environment that can be adopted by a Project and for that goal the author uses it to research the The Islamic Finance and Related Technology Strategic Vision in the Lebanese Context (IFSVLC). The IFSVLC is supported by a central Decision-Making System (DMS) for GG (DMS4GG), Organizational Engineering, and Enterprise Architecture (EA) based Projects. The Proof of Concept (PoC) is based on resources collected on the MEA and more specifically Lebanon; where the focus is on the FinTech, State Owned Global Financial Predators (SOGFP), coexistence and business collaboration between belligerent parties, to offer solutions. Such strategic Projects are managed by national Transformation Managers/Leaders (simply a Manager), like the late Rafic Hariri who was responsible for the implementation of a very complex Project, the reconstruction of Byrut-city and its Lebanese Financial System (LFS). Rafic Hariri (1944 – 2005) was the Prime Minister of Lebanon, and h headed five cabinets during his tenure and was widely credited for his role in constructing of the *Taif Agreement*. He played a crucial role in reconstructing of Byrut and LFS. He was assassinated in 2005 by the Iranian-militia members, were indicted for the assassination and were tried in absentia by the *Special Tribunal for Lebanon*. The IFRSVLC and related Project to become a part of a dynamic MEA with a modern GG Concept (GGC), business, marketing and a strategic financial transformational-plan.

A priority should be set to instore, intelligent GGC and security concepts that would be the skeleton of MEA's GGC, and used to integrate the global economy; and above all to detect and block SOGFPs' attempts of destabilizations and the ones inflected to Lebanon. An optimal example to mimic a GGC in Byrut's context, which is the reflection of all possible MEA's features, is to propose a generic transformational pattern. The Project for an organization, region, or country (or simply Entity) needs inter-community/religious (and ethnical), national-cultural, civics, business/marketing/financial engineering specificities/constraints to be mapped to GGC; and should include advanced ICS/FinTech which may support the Entity's finance and business longevity. The ongoing religious (and ethnical), financial and political MEA's crisis, make GGRisks high. To support a Project, Critical Success Areas (CSA) and CSFs must be used to mitigate GGRisks. The Project's CSFs can be configured to manage GCC's complexities and its uncertain evolution. Projects might involve innovation initiatives, like, the digitization processes, where automation enables AI-models optimize IFSVLC and GGC's processes. This Research and Development Project (RDP) background is related to Projects that use CSFs which are managed by the In-House Implemented (IHI) Polymathic Transformation Framework (IHIPTF) (Trad, & Kalpić, 2018a), which supports: 1) Polymathic-holistic approach; 2) GGRisks' mitigation; 3) Interfaces various APplication Domains (APD); 4) Considers religious/ethnical MEA specificities; 5) Analyses Moslem societal/cultural and financial CSFs; and 6) GGC's integration.

BACKGROUND

ICS/FinTech support the Project and its different types of GGCs and methodologies; and the common model to be supported, is the transactional economy model that generates revenue via the ICS and World Wide Web (WWW). Actual ICS/technology hype is shifting to Web 3.0 (Joseph, 2014). Major MEA's GG Events (GGEvent) are: 1) 9/11 terrorist act that has hit the societal, financial and cultural system; 2) Continuous Arab/Israeli wars; and 3) 4th of August (2020) Byrut's port-blast that devastated Lebanon's capital. This blast damaged thr LFS. Other chained GGEvents disrupted the West, mainly United States of America's (USA) unilateral domination of the MEA and endangered the equilibrium of the pro-western Financial-hubs. Tensions and the loss MEA's oil fields and financial/business markets are weakening the once robust western-model, and the main reasons are: 1) West's military degradation; 2) Short-sighted financial/commercial and cupid strategy; 3) Aggressive SOGFP, like the case of Heavens' plundering of Lebanon; 4) Rise of global competition like BRICS; and 4) Evident GG incapacities. These various GG-transformations causes continuous instabilities like the case of the Arab-spring, Euro-Centric Far Right (ECFR) and the emergence of Islamist-State (II). II caused wars, like the Syrian civil war (Trad, & Kalpić, 2019a). Another important fact, is the immense gas/oil fields, newly found in the east-Mediterranean, which cause tensions between Europe's southern states, USA, Russia, Iran, Israel and Turkey. As mentioned, the focus is on the MEA/Lebanon and the influence of FinTech and Islamic-finance.

THE RDP

Introduction

Figure 1. Project's levels' interactions (Trad, & Kalpić, 2017b, 2017c).

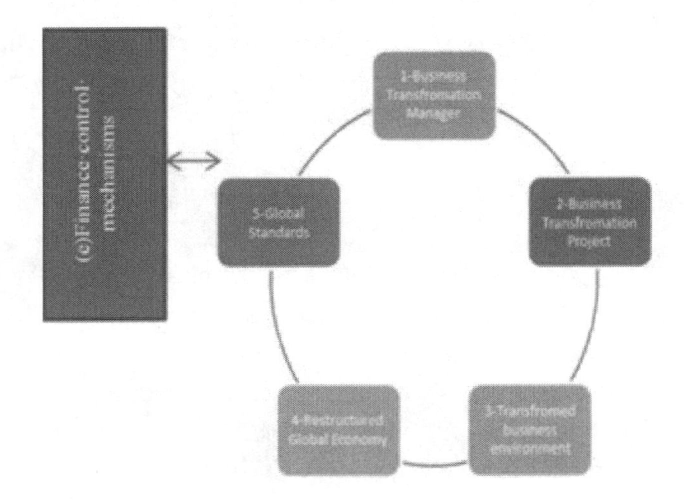

The IFSVLC tries to analyse how Islamic-finance (or sharia-compliant finance) which is a set of activities that comply with sharia (or the Islamic-law), can integrate complex Entities, like Lebanon. Islamic-finance include: El-Mudarabah (profit-sharing and loss-bearing), Wadiah (safekeeping), Musharaka (joint venture), Murabahah (cost-plus) and Ijara (leasing). The honorable Qur'an prohibits *el-riba*, which means increase in profit, which is a *haraam* (restricted or excluded). Modern Islamic finance is standardized, by organizations like the Islamic Financial Services Board (IFSB) and regulators in Bahrain, Indonesia, Jordan, Kuwait, Lebanon, Malaysia and Pakistan have developed guidelines for the related standards. Some countries, like Indonesia, Kuwait, Malaysia, Pakistan, Sudan, and the UAE have centralized these standards (Wikipedia, 2020a).

The Structure

The RDP is based on CSFs and a Literature Review Process (LRP) based on credible sources, like the Gartner Inc., a quantitative-qualitative mixed methodology and offers a set of recommendations to avoid fatal SOGFP scenarios that ruined many countries like the Lebanon, Iraq, France, and many other countries. XHFRs are due to the complexities encountered in the Project's implementation phase (Bruce, 1994). To enhance success-rates and to simplify GGC's integration, the author proposes the IHIPTF, which enables linking of the AHMM4GG to all levels of the IFSVLC, as shown in Figure 1 (Agievich, 2014). This chapter's RQ is: *Can an Islamic-Finance Strategic Vision for GGC be used in the MEA and more specifically in Lebanon?* The CSF-based RDP instance would use the DMS4GG and unbundling activities to implement the PoC (Trad & Kalpić, 2017b, 2017c).

Transforming Financial, Business and Marketing Activities

That is why the notions of EA/FinTech result from a *body of knowledge, principles…* address a crucial question: How does design-concepts facilitate IFSVLC's integration. But EA/FinTech alone cannot solve all problems related to Projects, therefore, EA/FinTech provides the ability to implement the coordination of: 1) Entities' processes and resources; 2) DMS4GG; 3) Project and GGC's interfacing, 4) IFSVLC's Integration; and 5) Predict SOGFPs' misdeeds. That needs qualified and ethical personal skills, that is a weakness, like in the case of Riad Salamé (Courrier International, 2024, OLJ, 2023; Trad, Nakitende, Oke, 2021).

The Role of Education and Ethics

The handicap in high-qualification jobs like in the Banque Du Liban (BDL) is the emigration of qualified and ethical specialists. SOGFP's plundering is a fatal concern, towards developing and maintaining viable LFS. The MEA and especially Lebanon, Manager*'s* selection process is very important for the Project, where his role of ethics and loyalty are of the utmost importance, and can also include: 1) Cultural/religious awareness; 2) Business and marketing integration; 3) Real-time accesses of EA/FinTech processes; and 4) Blocking SOGFP-mechanisms. The honest-technocrat's profile is recommended as a *base profile* for Project and its GGC Manager*s* (Trad, & Kalpić, 2018e).

RDP's Critical Success Factors

Figure 2. Average is 8.0.

Critical Success Factors	KPIs	Weightings
CSF_RDP_Basics_Structure	Feasible	From 1 to 10. 09 Selected
CSF_RDP_Transformation_Processes	VeryComplex	From 1 to 10. 07 Selected
CSF_RDP_IHIPTF	Feasible	From 1 to 10. 09 Selected
CSF_RDP_Education_Ethics	VeryComplex	From 1 to 10. 07 Selected

Based on the CSF review process, RDP's CSFs are presented in Figure 2, which is 8.0 (that is complex).

THE APPLIED CASE STUDIES

SOGFP Models

The MEA is rich with natural resources and this fact, attracts SOGFP-greediness and cupidity, like Known Safe Heavens (simply Heavens). Generated-conflicts have a usual-pattern for supporting massive SOGP-looting scenarios that can be described as a set of the following facts/actions (Courrier International, 2024, OLJ, 2023; Trad, Nakitende, Oke, 2021):

- SOGFP actors find possible natural (or financial) resources, like for example, oil/gas, financial assets, new markets…; and plan actions to grab and plunder. Like in many cases in the MEA and its surrounding region: 1) Iraq; 2) Lebanon; 3) Libya; 4) Kuwait; 5) France… The plundered values are astronomic and have levels of trillions…
- Localize richness in Entities, like in the MEA/Lebanon… Here the mighty SOGFP and associated banks information services network, apply the preparations for a plundering and the targeted Entity's destabilization offensive.
- Interact with actual super-power(s) and world elites to propose the SOGFP to solve world GGProblems; and of as usual divert massive capital to Heavens …
- Destabilize and eventually create armed-conflicts, like regional, civil or other types of war…
- Sell arms while the concerned countries subscribe huge debts in Heavens-banks… And these banks manage the arm-dealings processes…
- Manipulate currencies, so the debt for the Entities becomes huge…
- Convince the Elite (like BDL's Mr Salamé) of the embattled Entity to transfer the whole capital and assets to Heavens…
- Use psychology and active SOGFP-groups to discredit and even eliminate eventual belligerents…
- Finance and organize peace talks, to finance a settlement and apply of SOGFP-misdeeds.
- In many cases, the Entity's system changes which are used by SOGFPs seize the capital and assets…
- Many wealthy people, who invest in Heavens, with time, face major problems like illness, divorce, death… And this an ideal situation to apply SOGFP tactics.
- Ambush tactics, like the case of Polanski…

Modelling MEA's Case

IFSVLC uses the Applied Case Studies (ACS) for GG (ACS4GG) that include works, like Samir Kassir's that are fundamental to understand this chapter's (Kassir, 2010). The IFSVLC uses CSFs to estimate the success-rate of the Project and IFSVLC's integration (Trad, & Kalpić, 2018f). ACS4GG's categories and roles, are:

- Finance, governance, law and technology category or the ACS4GG_FinTech category.
- Economy, Growth or the ACS4GG_EcoSys category.
- Conflict, political, ethnical and other or the ACS4GG_EthnoPol category.
- Geopolitics, basics, analysis and transformation or the ACS4GG_GeoPol category.
- Demography, immigration, integration or the ACS4GG_DemoIm category.
- Natural resources, oil-Gaz, minerals or the ACS4GG_NatRes category.
- (In)dependence, external resources, man power, agriculture… ACS4GG_InpFact category.
- FinTech/Cyber, Internet related technologies … ACS4GG_IoT4GG category

During the LRP, the author selected a large set of possible ACS4GG related cases; and the some important ACS4GGs which are used in this chapter, are:

- Iranian influence case: in which the MEA is facing a delicate and turbulent period (in fact it also had), where today, a major obstacle for a lasting and sustainable peace, and evolution in the MEA is jeopardized by confrontational rivalries between the Iranian and the Western bloc (which includes countries opposed to it, like Saudi Arabia, the United Arab Emirates and Israel, and are strongly supported by the USA). Lebanon which is historically anchored to the West and to its historical ally France, holds a schizophrenic position, because of the (t) Iranic occupation, thru its proxy terrorist militias, like the Hezbollah. USA's administration confrontation with Iran is increasing MEA's tensions (Geranmayeh, 2018). And Iranian-militias occupy various MEA Entities (Agger, Jensen, 1996).

- Lebanese economy case: which analyses the social and economic repercussions of the Syrian crisis and civil war, on various Lebanese communities including the refugees' communities in Lebanon. These communities ignited a growing interest in understanding the CSFs that have implications on the growth of Micro, Small and Medium Enterprises (MSMEs) in Lebanon. The Lebanese government, donors and international banks (and institutions) have been increasingly assisting programmes for Lebanese MSMEs in their response to the ongoing fatal crisis. The support of MSME growth hinges on the overall objective of enhancing economic growth and creating jobs for the vulnerable Lebanese citizens and Syrian refugees. The vibrant MSME sector played a key role in creating jobs and maintaining economic growth. In turn, an in-depth analysis of the obstacles to the growth of MSMEs is necessary to evaluate how SMEs can overcome barriers to growth (Srour, & Chaaban, 2017).
- Lebanese resilience: where Lebanon has one of the most diversified economies in the MENA region, with growth traditionally driven by real estate, construction and tourism. With the services sector accounting for 76% of value added and manufacturing representing a high proportion of industrial production, Lebanon has a very developed economic structure (World Bank Indicators, 2017). Coupled with a high level of openness, a well-developed banking system and a strong private sector, the economic fundamentals position it well to resist to shocks. However, policy uncertainties and macroeconomic imbalances limit Lebanon's resilience. The Syrian crisis has strained Lebanon's public finances and service delivery. Poverty incidence among Lebanese citizens and income inequality have increased. Beyond the impact of the crisis, recent economic developments point to an underlying erosion of competitiveness and productivity levels. With an increasing fiscal deficit and high public debt, a significant fiscal adjustment is necessary, including restraining public wages, gradually reducing energy subsidies and increasing tax rates (MENA-OECD, 2018).
- The case of resources: a very vital sector, the security of energy supplies is crucial for any great power. Resources' security is not just a priority for major oil companies, it is a universal value. Therefore, when

the dangers rise exponentially, like in the MEA and north Africa, the world economic situation is jeopardized. This case's main topics are (Gomart, 2016):

1) Main businesses can be endangered or even stopped by the decisions of independent states. Powers cannot economize, therefore, tracking of their geopolitical environment becomes vital and their a need to react geopolitically.
2) Concerning diplomacy, countries like, Russia, China and the USA are permanent members of the United Nations Security Council; where they play a crucial role in all major international topics. From the military point of view, they are nuclear powers, in the same time, they are maintaining their formidable conventional weaponry and they occupy the top three positions in the world ranking of military spending.
3) Today, economically, China and the USA are considered as equals from the Gross Domestic Product (GDP) standards' point of view, while Russia falls far behind. Whereas, the USA remains, probably, the dominant power, China is emerging as the rising super power and Russia is for the moment a declining power.
4) To secure the trade of goods, these countries protect their traders and, by means of agreements between major powers, ensures the freedom and security of commercial exchange.
5) The global business and commercial world, may have difficulties in forecasting pitfalls of some foreign policies and, in extreme cases, of accepting the primacy of politics.
6) To understand the competitors, Entity's (societal or business) leaders must apply Gulf Cooperation Countries (GCC) to understand the historical and geopolitical CSFs, and in order to mitigate various types of GGRisks. GGRisks which may englobe, the trajectory of the target country and the related security challenges.
7) Since the end of the Cold War, six main CSAs have dominated Russian-USA relations: nuclear weapons, proliferation, the post-Soviet space, European security, the Arab world, and human rights.

8) Economic conditions have never figured high on the agenda, where the USA-Russian relationship seen a blow before the Ukraine crisis; when the USA admitted that the *reset* activated by president Obama had failed.
9) A group of USA's experts, is lobbying for a Russian-USA-Chinese alliance focused on ensuring nuclear security, the stabilization of the Moslem world and the stability of North Korea.
10) Revenues from energy resources were funneled into a security policy that had become essential in 1990s; and then, Russia was able to increase its military power. This new approach led to a renewed focus on Russia, given that its security is the one of its President.
11) This new evolution of extreme personalization of power may lead to different scenarios, depending on GGEvents linked to the next presidential elections (years 2018 and 2024). However modern and powerful it might pretend to be, today's Russia follows a geopolitical roadmap that points in the direction of resistance, confrontation and revanchism.
12) The Asia-Pacific region, is China's most important region, and the CSFs that are related to overlapping economic, energy resources and security policies. That implies, that China must work on its relations with all of its neighboring countries, not only using economic diplomacy and by applying soft power, but also by using coercive diplomacy and hard power, particularly in the regions East and South China seas.
13) Today, China can contribute directly to the global energy security stability policies, which are still dominated by the USA. China's GCC is driven by an optimistic diplomacy and by a strong leadership which tries to apply a grand strategy.
14) In the next period, or next decades, the decisions made by the USA and China will define the world's security and energy security policies. Despite ultra-fast transformations in the energy industry, the connection between oil supplies and national security remains firmly stable. The global policy of world energy and security has traversed different phases since 1945, and is today still the foundation of the world order and its stability; that was established by the USA and other world powers. The question today is: *Over what period of time will China try to change it?*

- The Merger/alliance-coalition, an Entity transformation process: this ACS4GG describes, a merger and alliances, like in the ACS4GG_FinTech, where three independent insurance companies and the transformed Entity consists of three divisions and headquarters. The ACS4GG_FinTech was designed to take advantage of synergies between three organizations (Jonkers, Band, & Quartel, 2012).
- Islamic finance case: describes an Entity which wants to acquire an asset and is composed of the following facts (Keraine, 2019):

    1) The Entity creates a local-mudaraba of which the Entity is an agent. The local-mudaraba issues bonds (or sukuk) that allows it to acquire the located property. The Entity agent's mandate is terminated and a deferred sales contract is set up in which the Entity acquires the property sold by local-mudaraba with a profit fixed from the outset. The sale price, which includes profit, is paid by the company to the local-mudaraba, according to a schedule listed in the contract. The money collected can be used by the Entity to reimburse the coupons to investors respected a schedule:
        1) The Entity is advised by the investment bank it has requested.
        2) An ad hoc local-mudaraba that was created by the Entity.
        3) The Sharia Committee analyses local-mudaraba's and sukuk's conformity.
        4) The local-mudaraba issues bonds (sukuk) to investors.
        5) Local-mudaraba collects funds from investors.
        6) The local-mudaraba pays the supplier the amount of the purchased good.
        7) The Entity's agent of the local-mudaraba, takes the delivery request of the property.
        8) The Entity acquires the property using local-mudaraba with a settlement, by the so-called maturity.
        9) The local-mudaraba pays investors (deducing the fees of the manager (mudarib) and the agent (wakil).

- The Beirut blast case: this major and historically recurrent catastrophe on the Lebanese population and its destiny, is a typical situation that has similar actions that have happened in the past. Lebanon's traditional enemies, have tried again, to assassinate Lebanon's hope... Happily again, the truthful ally, France, has ran to rescue Lebanon...

- The major SOGFP crime that was coordinated by major Lebanese personalities like Riad Salamé and Swiss banks that plundered and destroyed Lebanon (Courrier International, 2024 ; OLJ, 2023). It is exactly like in 1968, where Intra Bank's founder a SOGFP legend and the destroyer of the Lebanese financial establishment, Youssef Beidas, dies in Switzerland in particular circumstances. At the time of his death, he was classified by the Lebanese government as a notorious criminal and international terrorist. It is a recurrent model, how can it happen again? (Trad, Nakitende, Oke, 2021).
- These ACS4GGs propose sets of CSFs.

Integrating CSFs

This RDP focuses on the IFSVLC to assist GGC's integration, where the IFSVLC uses a Heuristics Decision Tree (HDT) that supports a wide class of GGProblem types, and it is a major benefit for Entities (Markides, 2011). The HDT based DMS4GG offers a set of solutions which are translated to recommendations. A CSF is measurable and is mapped to weightings that is estimated in the first GCC's iteration and then tuned through Architecture Development Method's (ADM) iterations, in order to verify IFSVLC's integration feasibility. A holistic GGC architecture and resultant sets of CSFs are essential for its functioning (Morrison, 2016). The main problem is, how to define what is really IFSVLC's influence on the Lebanon/MEA, and can it be a possible solution. Knowing that elite (Heavens) Entities have immense industrial/financial capabilities, but are SOGFPs; and Lebanon should avoid such Heavens, which are worldwide center of organized SOGFP misdeeds. So how can the MEA/Lebanon handle or adapt IFSVLC taking into account SOGFP activities. And knowing that at the moment of writing this chapter, Heavens have completely-plundered the whole of Lebanese wealth and the Lebanese are starving.

The Integration of Business, Financial and Legal Framework

Global financial conventions, laws and taxations, need and an IHIPTF, to support IFSVLC's integration; where Lebanon, after SOGFP's brutal misdeeds, has to transform its actual LFS and look at MEA's existing solutions. The notions of jurisdictions and regulations depend on the transaction's dispositions and the applied laws. FinTech forces MEA's Entities to implement global regulations and laws and avoid SOGFPs; and such SOGFPs or Heavens, impose regulation frameworks which themselves do not respect (Penn, & Arias, 2009).

ACS4GG's Critical Success Factors

Figure 3. ACS4GG's CSFs have an average of 7.25.

Critical Success Factors	KPIs	Weightings
CSF_ACS4GG_SOGFP_Models	VeryComplex	From 1 to 10. 07 Selected
CSF_ACS4GG_Modelling	VeryComplex	From 1 to 10. 07 Selected
CSF_ACS4GG_Integrating_CSFs	Complex	From 1 to 10. 08 Selected
CSF_ACS4GG_Integration_Financial_Framework	VeryComplex	From 1 to 10. 07 Selected

ACS4GG's CSFs are used and evaluated and are represented in Figure 3, and the result is 7.25 (complex).

AHMM4GG'S USAGE

Basic Elements and the MM

IFSVLC's CSFs support HDT's algorithm-nodes that are identified as vital for successful targets to be reached and maintained. An HDT node is AHMM4GG's basic element that is needed for the Projects to estimate its success-rate; using the exposed two phases: 1) Feasibility check; and 2) GG-Problem solving) (Morrison, 2016). The IFSVLC uses AHMM4GG that is an abstract model containing a proprietary Natural Language Programming

(NLP) for GG (NLP4GG) that can be used to implement the IFSVLC/GGC. The AHMM4GG nomenclature is presented to the reader in Figure 4.

Figure 4. AHMM4GG's nomenclature (Trad, & Kalpić, 2020a).

Basic Mathematical Model's (BMM) Nomenclature

Iteration	= An integer variable "*i*" that denotes a *Project/ADM iteration*	
microRequirement	= (maps to) KPI	(N1)
CSF	= Σ KPI	(N2)
Requirement	= (maps to) CSF = $\bigcup$ microRequirement	(N3)
CSA	= Σ CSF	(N4)
microMapping microArtefact/Req	= microArtefact + (maps to) microRequirement	(N5)
microKnowledgeArtefact	= $\bigcup$ knowledgeItem(s)	(N6)
neuron	= action->data + microKnowledgeArtefact	(N7)
microArtefact / neural network	= $\bigcup$ neurons	(N8)
microArtefactScenario	= $\bigcup$ microartefact	(N9)
AI/Decision Making	= $\bigcup$ microArtefactScenario	(N10)
microEntity	= $\bigcup$ microArtefact	(N11)
Entity or Enterprise	= $\bigcup$ microEntity	(N12)
EnityIntelligence	= $\bigcup$ AI/Decision Making	(N13)
BMM(*Iteration*) as an instance	= EnityIntelligence(*Iteration*)	(N14)

The Generic AHMM's Formulation

AHMM	= $\bigcup$ ADMs + BMMs	(N15)

AHMM's Application and Instantiation for GG (and IFSVLC)

Domain	= GG	(N16)
AHMM4(*Domain*)	= $\bigcup$ ADMs + BMMs(*Domain*)	(N17)

The IFSVLC Transformation Mathematical Model

The GCC combines Project methodologies, and the AHMM4GG that integrates the Entity's organisational-concept, and ICS/FinTech (Lazar, Motogna, & Parv, 2010). As shown in Figure 5, the AHMM4GG is a part and is the skeleton of the IHIPTF that uses NLP4GG scripts to support GGC and IFSVLC requests. IFSVLC's components interface the DMS4GG and

Knowledge Management System (KMS) for GG (KMS4GG) (simply Intelligence), as shown in Figure 3, to evaluate/manage and map CSFs. In phase 1, if the aggregations of all Project's CSAs/CSFs tables (CSA_DT) exceed the defined minimum the Projects continues to phase 2 (or the second part), the PoC, in which a concrete GGProblem is solved.

Figure 5. IHIPTF's components.

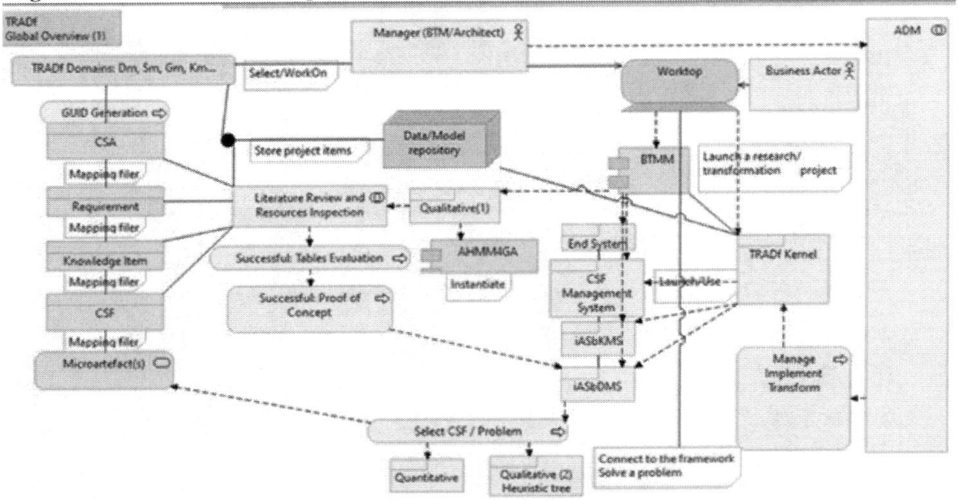

AHMM4GG's Critical Success Factors

Figure 6. AHMM4GG CSFs have an average of 8.0.

Critical Success Factors	KPIs	Weightings
CSF_AHMM4GG_Basic_Elements_MM	Possible	From 1 to 10. **09 Selected**
CSF_AHMM4GG_CSFs	Possible	From 1 to 10. **09 Selected**
CSF_AHMM4GG_HDT	VeryComplex	From 1 to 10. **07 Selected**
CSF_AHMM4GG_IFSVLC_Transformation_MM	VeryComplex	From 1 to 10. **07 Selected**

As shown in Figure 6, the result is 8.0 (Complex).

FINTECH AND ICS' INTEGRATION

FinTech's Integration

ICS/FinTech's landscape is based on the following facts (Kagan, 2020; Nakamoto, 2008):

- In 2016, FinTech start-ups received $17.4 billion in funding, where the most successful ones are today valued at $83.8 billion and will be valued to about $147.37 billion by the end of year 2018.
- The USA, produces far most of FinTech start-ups, with Asia as second. Global FinTech funding hits a new high in the first quarter of the year 2018 let by a significant uptick in deals in USA.
- Automated/Integrated Transactions or (e)coins, are a sequence of digital specific signatures, where each coin's owner transfers his (e)coin to the next node by electronically signing a hash key and linking it to the previous transaction and the public key of the next coin's owner; and then adding them to the end of the (e)coin transaction.

FinTech's most innovative areas are:

1) Cryptocurrency as digital cash.
2) Blockchain technology, including Ethereum, a Distributed Ledger Technology (DLT), which maintains records on ICS' network which has no central ledger.
3) Smart contracts, utilizes FinTech- blockchain programs to automate contracts conclusion between partners.
4) Open banking, which is based on blockchain and enables third-parties should have access to bank data to implement FinTech applications; in order to create a connected network of FinTech institutions and third-party partners.
5) Insurtech, where FinTech is used to simplify and streamline the insurance industry.
6) Regtech, to support FinTech companies to respect compliance rules, especially the ones covering Anti-Money Laundering and Know Your Customer protocols, which fights fraud.

7) Robo-advisors, uses algorithms to automate investment advice and minimizes costs and increases accessibility.
8) Unbanked/underbanked, services to support disadvantaged or low-income individuals who are ignored by banks or other financial services companies.
9) Cybersecurity, used to combat Cybercrime and to protect data. Cybersecurity and FinTech are intertwined.

FinTech Strategy

GG's uncertainty or crisis make FinTech investments decrease, but FinTech-ecosystem continues to evolve, which results in increasing the focus on possibilities, and major predictions are (KPMG, 2020):

- Bigger and bolder deals: where deal sizes will grow as investors focus on late-stage FinTechs.
- Product expansion: maturing FinTechs and competitive banks will continue to expand.
- Deals occurring in various locations: In which FinTech deals in jurisdictions outside traditional financial markets, such as in Southeast Asia, Latin America and Africa.
- Rise of BigTech: like Alibaba, Alphabet, Apple and Tencent will increase their efforts on FinTech's expansion, to englobe and control developing markets.
- Digital banking licenses: following the lead of Hong Kong (SAR) and Australia and Singapore, more other *Entities* in the Asia Pacific region will also develop digital banking systems and use digital banking licenses to improve competition and deliver services.
- The hunted start hunting, avant-garde FinTechs will increasingly make their own investments in other emerging FinTechs.
- Partnerships, will accelerate activities between major players.
- Open banking to open finance, focuses on open data possibilities which moves beyond banking.
- Re-bundling of financial services, includes the unbundling of FinTech products, which begins with the transformation of an Entity.
- Cybersecurity and digital identity management, will become essential for Entities.

Entities should protect themselves from so called Big Techs for locked-in situations and/or SOGFPs' misdeeds.

FinTech's Legal Framework

The International Organization of Securities Commissions (IOSCO) identified 8 areas that constitute FinTech, which are payments, insurance, planning, trading and investments, blockchain, lending/crowdfunding, data and analytics, and security. The growth of FinTech-market implies relevant issues and legal GGRisks. It is important that financial-regulations are simplified and to comply with regulations in many jurisdictions. The challenge for regulators is to find the right-balance between encouragement of the emerging ICS/FinTech and the regulatory-frameworks. Various legal issues were identified and must be considered when dealing with FinTech, which are based on the European Banking Authority's report, *Automatic report on the prudential GGRisks and opportunities arising for institutions from FinTech*, which can include: 1) Data protection and Cybersecurity; 2) DLT and smart contracts; 3) DLT is an amount of shared and synchronized data; 4) Robo-advisors and legal responsibility; 5) Outsourcing core banking/ payment system; and 6) Biometric authentications.

Bitcoin Block-chain Automation-Transactions

Peer-to-peer (e)cash supports FinTech's (e)transactions for parties from different financial institutions. (e)Signatures secures (e)transactions and provide a block-chain solution. The solution to costs is to use peer-to-peer communication, that uses timestamps (e)transactions by hashing them into an ongoing block-chain of the hash-based subsystem, that creates logging records. The communication infrastructure requires minimal supplementary resources. (e)Commerce relies practically on financial-institutions that support (e)payments. FinTech-systems work for all (e)transactions, but unfortunately there are inherent weaknesses of the infrastructure. It is not possible to achieve hundred percent security of non-reversible (e)transactions, because financial institutions cannot function without mediating conflicts.

Standards, Processes and Models

The LFS depends on Byrut and its Future Intelligent City Concept (FICC) based Projects to integrate i urban and automation-models that includes various APDs, and for creating new-values, as shown in Figure 4 (Kolbe, 2015). Standards like City Geography Markup Language (CityGML) can be used to automate and integrate other MEA cities and their Financial-hubs. The IFSVLC offers models that support Projects and Intelligence to: 1) Predict and avoid possible GGRisks, like SOGFP scenarios; 2) Locate SOGFP origins; and 3) To ensure sustainable benefits. To manage FinTech, WWW and FICC based Projects, an adequate mapping concept is used to integrate societal, governance, automation and technology standards (OASIS, 2014).

Figure 7. GA transformation urban model (Kolbe, 2015).

Linking Sectors creates a Lattice of Models

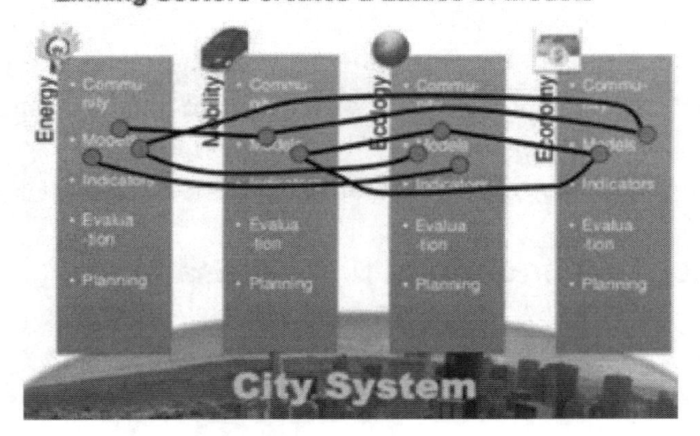

The principle and strategy impose a linear "1:1" mapping and standards having the levels (Lea, 2017): 1) Strategic; 2) Process; and 3) Technical; as shown in Figure 8. These standards and tooling environments support Projects through an iterative pseudo-bottom-up approach. For GGC's integration, distributed communication standards like the Internet of Things for GG (IoT4GG) can be used, in domains like, financial control, transportation, healthcare and many others… The mentioned domains are transformed in order to be used for the transition of old city structures to smart city, by applying common architecture elements supporting and controlling complex domains, like finance (BSI, 2015).

Figure 8. The standards' levels (Lea, 2017)

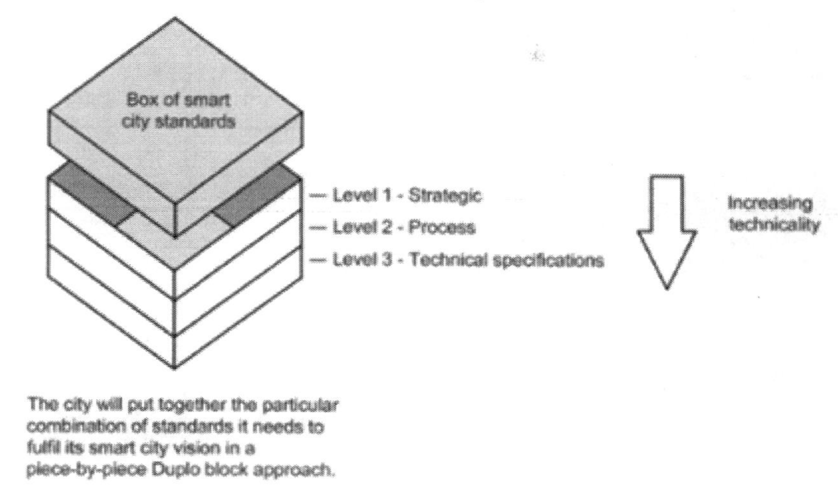

Electronic Payments and Finance

The IFSVLC use ICS/technologies to transform the quality-of-life where automating Financial-hubs, economy and digitalization of services are fundamental for transforming customers' behaviour. The evolution of technologies-based services is un-linear in different world-wide. Projects need to transform banks in order to automate crucial financial-services, like the case of the Deutsche Bank, Raiffeisen Bank, Hana-Bank and Bank Group as shown in Figure 9 (Makarchenko, Nerkararian, & Shmeleva, 2016).

Figure 9. Financial-services integration (Makarchenko, Nerkararian, & Shmeleva, 2016).

Banking is changing … with or without the banks. Response to the millennials. Financial services. (2015) http://oracledigitalbank.com/resources/pdf/DBOF_Industry_Research_Report.pdf

	Important	Current capability	Market lag
Mobile device payments	94%	44%	-50%
Market Lag	92%	24%	
Real-time analytics	90%	30%	-60%
Digital advisory service	83%	28%	-55%
Location-driven services	82%	19%	-63%
Offers via social media	78%	34%	-44%
Comparison services based on financial profile	76%	28%	-48%
Social media account management	72%	14%	-58%
Gamification	72%	15%	-57%
Digital personal assistant	67%	12%	-55%

Smart-devices are needed to interface Finance-data and functions; and can localize dangerous SOGFP activities. The evolution of smart scenarios and applications make lives superficially easier. IoT4GG based infrastructure offers scalable solutions (Miori, & Russo, 2014).

Reference EA-Models and its Phases

IFSVLC's reference EA-models support Project's implementation and is vendor-agnostic. EA-models use GGC standards that include various frameworks, like, EA/TOGAF, microartefacts interoperability… The proposed IHIPTF's EA capabilities, can help in establishing architectural guideline that defines the Project's initial phase and vision (The Open Group, 2011a, 2011b).

Figure 10. EA's implementation spectrum (Burton, & Burke, 2012).

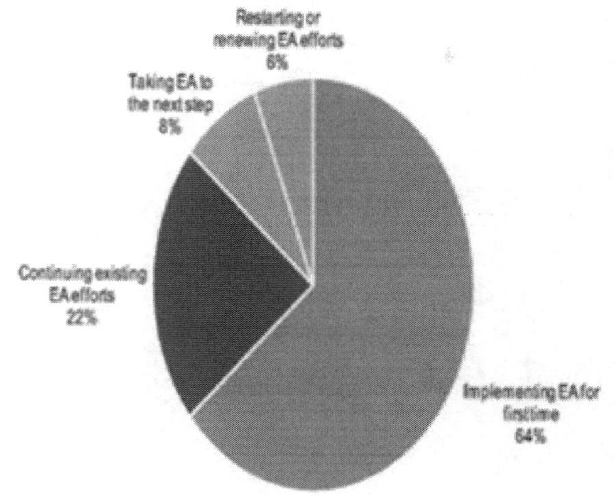

MEA's GGC integrators have a high-degree of focus on Entity's EA strategies and a lower-degree of importance to physical implementations; that causes GGProblems. Therefore, MEA's Entities have to develop local Project/EA skills for implementing GGCs and that delivers IHI EA models that focus on FinTech. Unfortunately, there is a lack of financial, business/ marketing regional collaboration and engagement. In a Gartner's survey on EA practitioners in the MEA, 64% of the respondents confirm that they are implementing EA for the first time, as shown in Figure 10. And it is actually a greenfield for GGC architecture integration. Surveyed EAs stated (with 88%, as shown in Figure 14) that their primary focus today was on aligning Entity's business vision, ICS and information technology to deliver a holistic strategic value for enabling major Project (Burton, & Burke, 2012).

Figure 11. Enterprise architecture's alignment survey (Burton, & Burke, 2012).

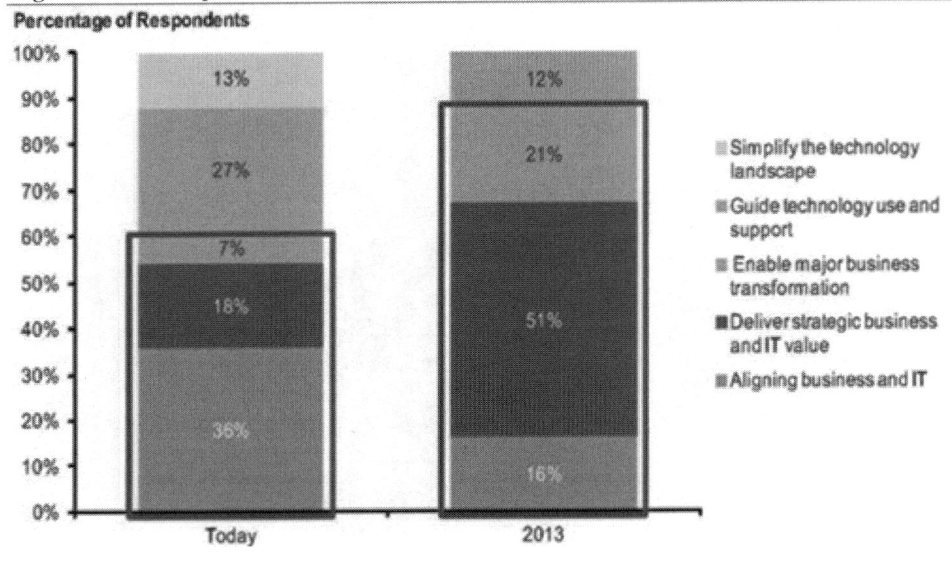

ICS's Critical Success Factors

Figure 12. ICS' CSFs have an average of 7,0.

Critical Success Factors	KPIs	Weightings
CSF_FinTech_ICS_Integration_Strategy	VeryComplex	From 1 to 10. 07 Selected
CSF_FinTech_ICS_Legal_Framework	VeryComplex	From 1 to 10. 07 Selected
CSF_FinTech_ICS_Bitcoin_Block-chain_ePayments	VeryComplex	From 1 to 10. 07 Selected
CSF_FinTech_ICS_EA-Models_Standards	VeryComplex	From 1 to 10. 07 Selected

As shown in Figure 12, the result is 7.0 (VeryComplex).

INTELLIGENCE'S INTEGRATION

Basics and Structure

Integration of AI in IFSVLC requires or supports (Kismawadi, Aditchere, & Libeesh, 2024; Hamadou, Yumna, Hamadou, Jallow, 2024; Kismawadi, Irfan, Abdul, & Shah, 2023):

- GGRisks' management for sustainable development, financial stability, transparency, and Shariah compliance.
- Aid in expanding financial inclusion and achieving defined objectives.
- Unleashing AI's power like in the case of Bank Syariah Indonesia (BSI).
- Customers are cultured and well-informed and they prefer limiting human-interactions, so there is the need for automation.
- Improves the overall Financial-system's performance by providing automation and accurate results.
- Gained acceptance and important consideration worldwide.
- Related effects are enormous, and cannot be denied, its application is economically feasible and precise.
- Presents significant opportunities and obstacles like adherence to Islamic law, ethical concerns, risk mitigation…
- Revolutionizes Islamic-finance by increasing efficiency, decreasing costs, enhancing risk management, and facilitating personalized services.
- Challenges can also be job shifts, bias in decision-making, data-privacy, and security.
- Financial-products can violate Islamic-laws or engage in illicit behaviour.
- Ensures conducting ethically and responsibly, in accordance with Islamic-law.

Collaboration

For collaborative Intelligence the goal is to obtain the basis for managing information-items using the GGC; where items are associated with the corresponding CSFs. In the MEA many traditional knowledge management accesses exist and are a barrier for the implementation of an innovative Intelligence (Malhotra, 2005). The Intelligence manages electronic GG Intelligence Items (GGItem); where GGItems and related scripts are responsible for the manipulation of AI and control various Entity's Intelligence processes. Intelligence contains identifiable GGItems that link GGC objects using a unique identifier. GGItems map to CSFs and microartefact(s) and are classified in specific CSAs. A IFSVLC concept expresses a fundamental structural concept for the GGItem based implementation.

CSFs and Objectives

Figure 13. AI-Fields/Intelligence.

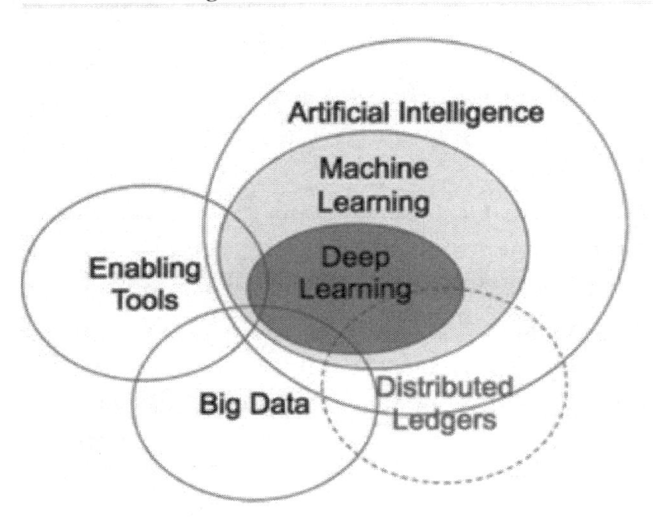

There is a solid focus of MEA's Projects on investing efforts in FinTech to deliver agile financial/business environments. There is a large gap among the top priorities of MEA Entities to align finance/business and ICS/Fintech to support Projects, because it is difficult to engage MEA's business and marketing Managers/Leaders in complex/risky Projects, which must prove their credibility by using AI/Intelligence (Burton, & Burke, 2012). Intelligence uses CSFs can be applied for the: 1) Selection of Project's resources; 2) Financial/business and marketing strategies definitions; 3) Support of HDT based engine; and 4) Training needs. Where Projects, have become in many Entities, major priorities and with previewed budgets of about $135 billion in 2021. In which the main objectives would be: 1) Traceability and protection against SOGFPs; 2) Livability; 3) Workability; and 4) Sustainability. As shown in Figure 9, various AI-Fields/Intelligence have enabled the integration of FinTech.

A Risky Transformation

Projects are very risky and GGRisks levels are very-high, and have 95% of XHFRs. Projects are even more risky, when using radical and long-term Projects. IHIPTF integrates GGRisks' management and performs GCCs by using Intelligence. In the context of the IFSVLC that is supported by the Intelligence, Projects' team- members can select and tune CSAs and their CSFs. The selected CSFs are orchestrated by AHMM4GG's choreography engine that is the base of the Intelligence (Hussain, Dillon, Chang, & Hussain, 2010; Trad, & Kalpić, 2017b, 2017c).

Intelligence's Critical Success Factors

Figure 14. Intelligence CSFs have an average of 7.75.

Critical Success Factors	KPIs	Weightings
CSF_AI_Intelligence_Basics_Structure	Possible	From 1 to 10. **09 Selected**
CSF_AI_Intelligence_Collaboration	Complex	From 1 to 10. **08 Selected**
CSF_AI_Intelligence_CSFs_Objectives	VeryComplex	From 1 to 10. **07 Selected**
CSF_AI_Intelligence_Risky_Transformation	VeryComplex	From 1 to 10. **07 Selected**

As shown in Figure 14, the result is 7.75 (Complex).

IFSVLC'S INTEGRATION

The Role and Risks of Intermediaries

Entities standardize their audit, governance, control and monitoring environments and EA-models. Intermediaries cause major SOGFP misdeeds and this proves that the IHIPTF/GGC are crucial for Entities and their Financial-hubs for their availability and tracing SOGFP misdeeds (Kowall & Fletcher, 2013). An ethical Manager should be a member of the executive strategy team, and should diligently work with the GGRisks and legal teams, where he can bring in an effective view to changes in the financial vision on engineering GGRisks and legal regulations, control and governance issues. These facts reduce the number of intermediaries, especially SOGFP ones. Financial intermediaries can be: Banks, Insurance companies, Mutual funds, Non-Banking Financial institutions, high-profile Managers like Mr Riad Salamé (Courrier International, 2024, OLJ, 2023) or Youssef Beidas (Trad, Nakitende, Oke, 2021) who destroyed the LFS by sending Lebanon's colossal assets to Heavens like Switzerland, and were convicted (OLJ, 2024).; or other...

The Role of Financial Services

Financial models support financial-services over FinTech based online-frontends. Projects can help banks and financial institutions, reshape their finance/business programmes, like after the 2008 financial crisis, where the IHIPTF/GGC focus was on GG's reach, redefining global initiatives and re-organizing trade-services and prioritizing a customer-centric approach, like in the case of Citibank. Citibank restructured its activity centres and redefined its global strategy to assist customers in solving financial-problems. Citibank's Project was quick and the landscape for treasury and trade-services was radically modified to stand-up again after the financial crisis. Multi-functional banking is a form of block-chaining, based on web-technologies that enabled low-cost and efficient financial-operations. Citibank was the first financial institution to finalize a Project and had a significant impact on the finance-business transformation models (Farhoomand & Lentini, 2008). In the case of the practically destroyed LFS, what would be the influence of IFSVLC.

The Influence of IFSVLC on the LFS

In 1954 to 1956, Raymond Eddé sponsored LFS reforms related to rent laws and banking, and founded the basis for LFS' confidential banking system that proved to be a factor in Lebanon's exponential economic growth in the following decades; but also, its debacle (Wikipedia, 2024a). Could a version of IFSVLC be an option? The Islamic-Finance industry has been growing at a rate of 15 to 20% in the past decades. The LFS and BDL, promulgated several laws and circulars to regulate this industry in 2004. But, the market's volume of Islamic-Financial operations was not sufficient, as a quantitative study indicates. The study suggests the application of Islamic-Financial instruments to securitization, privatization, SMEs and agricultural support, microfinance and Islamic-funds. That would boast the volume of operations of Islamic-financial institutions in the LFS (Chammas, 2006). Islamic-banking is an important topics, like financing economic development, identifying obstacles and studying ways to eliminate them… For the LFS the subjects like customers attitudes towards Islamic-Banking and the effect of Islamic-Banking on national economy, were important. But there are various obstacles for Islamic-Banking in Lebanon, that mainly due to local mentalities…

Knowing that already in 1963, attempts have been made to avoid riba-based banking transactions by establishing banks that operate in accordance with Islamic-Shari'a. That was followed by the establishment of the first Islamic-bank in Dubai in 1975, and succeeded in doing operations that comply with the Islamic-Shariah. This was followed by the establishment of the Islamic Development Bank, the Faisal Islamic Bank of Sudan, the Kuwait Finance House, Faisal Islamic Bank of Egypt, Dallah Al Baraka Bank, and other Islamic banks which spread across all five continents, reaching 450 banks in more than 75 countries. Today, the volume of Islamic-banking assets reaches about $1.7 Trillions, and various conventional-banks have branches of Islamic-banks. So, Islamic-banking (and finance) is an important factor in the Entity's economic and social development. The LFS must look deeper in Islamic-banks to examine their future in Lebanon, to analyse customers' attitudes, and have the adequate infrastructure (Sujud, & Hachem, 2018).

FinTech's Infrastructure

Gartner's 2014 Chief Information Officers (CIO) survey results showed that taking account of digital opportunities and threats that pervade every aspect of finance, business and government, the ICS as defined in a strategy of each MEA's Entity, various sectors like, industry and government is becoming unique, as shown in Figure 15.

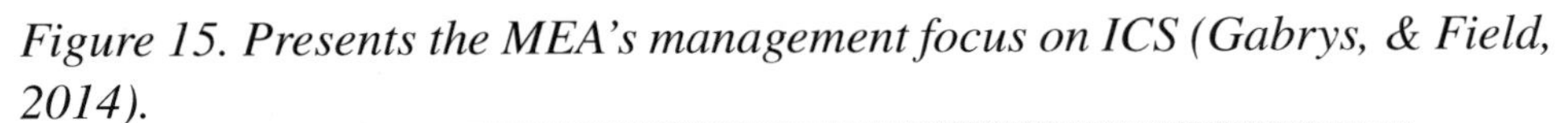

Figure 15. Presents the MEA's management focus on ICS (Gabrys, & Field, 2014).

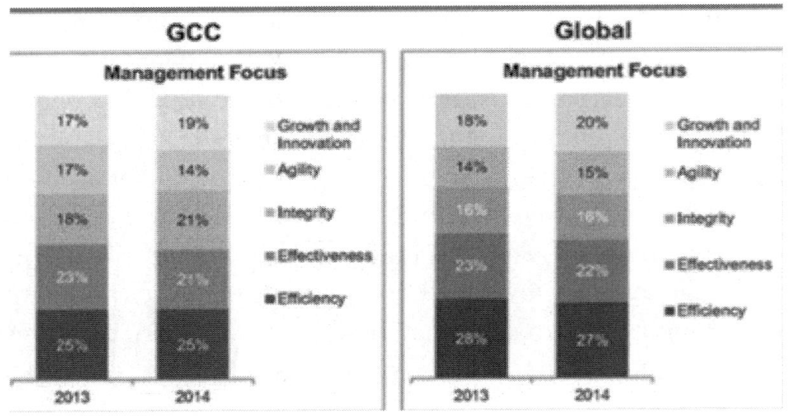

This RDP explores CIO's strategies for the MEA where 40% of companies (versus 25% globally) have adopted the distributed complex ICS infrastructures, as shown in Figure 16.

Figure 16. MEA's ICS sections (Gabrys, & Field, 2014).

	GCC		Global
1	BI/Analytics	1	BI/Analytics
2	Cloud	2	Infrastructure and Data Center
3	Infrastructure and Data Center	3	Mobile
4	Mobile	4	ERP
5	ERP	5	Cloud
6	Security	6	Networking, Voice and Data Communications
7	Customer Relationship Management	7	Digitalization/Digital Marketing
8	Networking, Voice and Data Communications	8	Security
9	Continuity	9	Industry-Specific Applications
10	Industry-Specific Applications	10	Customer Relationship Management
11	Application Development	11	Legacy Modernization
12	Architecture	12	Collaboration

Islamic Finance's Fundaments

Islamic -inance originated in the 7th century AD, with the Islamic conquest and it was revived in the period between 1950's to 60's with two main GGEvents (Keraine, 2019):

- 1956: Pilgrims' Administration and Fund (Tabung Hadjji) in Malaysia, that originated and was funded by the Malaysian government, and was dedicated to invest the capital collected from an important number of small savers in large Projects.
- 1963: MitGhamr's experience in Egypt, which was an entirely private Project, which collected savings and investments for cooperatives operating in the agricultural areas of northern Egypt. The goal, was to mediate financial resources between savers and small local investors.

Islamic finance refers to (Ross, 2020):

- The way, businesses and people raise and manage capital in accordance with the Sharia, or the Islamic law.
- It also refers to various types of investments that are considered permissible under this form of the Sharia law.
- It can be seen as a unique form of an Entity's societal responsible for investment schemes.
- Concerning its subbranches, the burgeoning field is strictly scrutinized.

Islamic Banking and Taxation

Islamic finance started in the 7th century AD, but it has been gradually formalized since the 1960s. This formalization process was inspired by the tremendous oil wealthy Entities, which encouraged and reinitiated the interest and demand for Sharia compatible financial and economic products and services. Where the concept of GGRisk sharing in raising capital, is an important CSF for Islamic banking and finance. Islamic finance constraints are applied to avoid the cases of: 1) The riba (usury); and 2) The gharar (ambiguity or deception). Islamic banks are available and operational in Saudi Arabia, Qatar and Oman. GCC represent about 90 percent of the total assets of Islamic banks in Arab countries. Where nearly half of these

Sharia-compliant banks are in Saudi Arabia, followed by the UAE with 20 percent, Kuwait with 17.4 percent and Bahrain with 11 percent (Albawaba Business, 2014). The list of Islamic banks in Lebanon are: 1) Arab Finance Investment House; 2) Al Baraka Bank; 3) Arab Banking Corporation; 3) Gulf International Bank; 4) Lebanese Islamic Bank; 5) Byblos Bank Africa Ltd; 6) Gulf International Bank; and 7) National Bank of Kuwait. Islamic-banks are subject to double taxation of their operations because they are merchants which acquire assets before transactions' beginnings. Moreover, structuring Sharia-compliant products into separate contracts, needs financial stamping duties. These double fees, value added taxes and stamp duties represent a major burden and increase the cost of financing (Albawaba Business, 2014).

Islamic Law

Islamic law's main facts are (Ross, 2020):

- Views lending with interest reimbursements as a relationship that favours the lender, who dictates charges interest at the borrower's expense.
- Considers money as a measuring instrument for value and not an asset, therefore, a person should not be able to receive income from money alone.
- Considers that interest is deemed riba and such practice is proscribed; or it is *haram*, which means prohibited. It is treated as usurious and exploitative. By contrast, Islamic banking exists to guarantee societal-geoeconomical goals of Islamic Entities.
- Sharia-compliant finance (halal, which means permitted), consists of banking and financial institution shares, in sense of Entity's profit or loss and it underwrites them.
- The important concept of gharar, in a financial context, refers to the ambiguity and deception that come from transactions whose existence and value are uncertain. Concrete, financial domains of gharar are the insurance industries.
- Equity financing of Entities is permissible, as far as those Entities are not engaged in restricted businesses. Prohibited or haram activities, include producing alcohol, gambling and pornography products.

Financial Agreements

Permissible financial agreements in Islamic banking and finance, are of the following types:

- Profit and loss sharing contracts (Mudarabah), where Islamic bank investors, assume a share of the profits and losses. This process is agreed with the depositors. Islamic banks, filter *Entities* balance sheets to determine whether Entities' sources of income are prohibited. Entities having important debts or engaged in suspicious businesses are excluded.
- Declining balance shared equity: calls for the bank and the investor to purchase the product together, like for example, purchasing a house. The bank gradually transfers its equity, in the house to the individual homeowner, whose payments constitute the homeowner's equity.
- Lease to own: is similar to the declining balance of the previous point, except that the bank puts practically all the money of the house's value. A portion of every payment goes toward the lease and the balance towards the home's initial purchase price.
- Instalment sale (Murabaha): starts with an intermediary transaction with a free and clear label. The intermediary investor, then agrees on a transaction's conditions with the buyer; this transaction includes some profit; using a series of deferred (instalment) payments. This credit transaction is an acceptable form of exchange and should not be compared with an interest-bearing loan.
- Leasing (Ijarah): is a transaction that involves having the right to use an object (usufruct) for limited time, with the condition, that the lessor must own the leased object for the duration of the lease.
- Islamic forwards (Salam and Istisna): are transactions for certain types of business and is an exception to gharar. A prepaid approach, where the item is delivered at a definite point in the future.
- Basic investment vehicles, with permissible Islamic investments that are listed below:

  1) Equities: Sharia law allows investment in company shares (common stock) as long as those companies do not engage in forbidden activities. Islamic scholars have made some concessions on permissible

companies, as most use debt either to address liquidity shortages (they borrow) or to invest excess cash (interest-bearing instruments).

2) Fixed-Income: Retirees who want their investments to comply with the tenets of Islam face a dilemma in that fixed-income investments include riba, which is forbidden. Therefore, specific types of investment in real estate could provide steady retirement income while not violating Sharia laws. These investments can be direct or securitized, like, diversified real estate fund.
3) Leasing bond-equivalent (ijarah sukuk): where the issuer sells the financial certificates to an investor group. The group will own the certificates before renting them back to the issuer in exchange for a predetermined rental return. As with the interest rate on a conventional bond, the rental return may be a fixed or floating rate pegged to a benchmark, such as the LIBOR. The issuer makes a binding promise to buy back the bonds at a future date at convened value. Special Purpose Vehicles (SPV) are often set up to act as intermediaries in financial transactions.

Profit and Loss Sharing

There are two major profit and loss sharing equity-based models of financing in Islamic finance (Mudharaba and Musharaka) (Albawaba Business, 2014):

- Musharaka is the establishment of a partnership or joint venture between the bank and the client for a specific financial project. It is a *profit and loss sharing* financing agreement, in which the bank and its client both invest their capital and expertise to the concluded deal.
- Mudharaba, as in Musharaka financing, takes the form of a partnership or joint venture between the bank or investor (Rab al-Mal) and the client or entrepreneur (Mudarib), for a specific financial project. However, the main difference is that the bank provides the capital while the client provides the expertise, skills and management.

Islamic Insurance

Traditional insurance is forbidden as a tool for GGRisk management in Islamic law, because it constitutes a transaction with an uncertain outcome (a form of gharar). Insurers use fixed income (a type of riba), as a part of their portfolio management process to satisfy liabilities. A possibility for a Sharia compatible offer is cooperative (mutual) insurance. Clients contribute to a common fund, which is invested in a Sharia-compliant system. Funds are used to satisfy claims and unclaimed profits are distributed among clients (Ross, 2020).

Islamic Finance in the GGC

Islamic finance is gaining worldwide recognition, because of its ethical and economic principles. The increasing development of Moslem *Entities*, this field will witness rapid evolution. Islamic finance will continue to address the issues of integrating Islamic investment policies and modern portfolio theory (Ross, 2020).

Islamic Finance in Mixed Entities

In the MEA, there are countries with important non-Moslem communities like in Lebanon, where Islamic finance, faced the following facts and challenges (Albawaba Business, 2014):

- Years after the introduction of Islamic banks in Lebanon and despite efforts by the Lebanese Central Bank to regulate Islamic finance, Sharia-compliant lenders have to adapt to the Lebanese complex banking sector. Central Bank's statistics show that Islamic banks in Lebanon had $674 million in assets in 2012 and that in 2013 the amount had only grown to $712 million. This included four Sharia-compliant banks operating in Lebanon, which are failing to conquer local market share. And their share is not beyond one percent.
- Bankers and financial analysts interviewed by *The Daily Star* gave, concerning Islamic finance, said that they believed that it failed to make a mark in the Lebanese banking sector because of the lack of awareness of the various Islamic banking options. This fact was con-

firmed by Adnan Youssef, president and CEO of Al-Baraka Banking Group. Youssef, admitted that such a process would take time whereas in the Gulf region, Islamic finance is booming.

- Ghassan Chammas, adviser to the board of directors of BLOM Development Bank, describes it as a black box… And that it is a costly process; added to that, fiscal laws resisted to Islamic banking operations in Lebanon.
- Rima Turk Ariss, a finance professor at the Lebanese American University (LAU), argues that Islamic banking in Lebanon focuses on debt-based contracts, known as Murabaha and not on equity-based contracts; and these two approaches are hard to integrate.
- Raed Charafeddine, the Central Bank's first vice governor, argues that Sharia-compliant banking's unpopularity in Lebanon, is because clients are not drawn to non-guaranteed deposits.

The Lebanese Legal System

The Lebanese Legal System (LLS) and Islamic finance's integration face the following hurdles (Chedid, 2018):

- LLS has implemented a pluralistic approach for banking that provides for both Islamic and conventional banks' well-defined integration processes. However, despite a complete regulatory framework, Islamic banks still operate in the context dominated by conventional banking.
- Law No 575/2004 related articles, provide a legal framework for Islamic finance pursuant to which the *Banque du Liban* (BDL) and the Capital Markets Authority (CMA) have issued corresponding regulations.
- Upon its establishment, the CMA took up the main provisions of BDL's decision on Islamic collective investment schemes (CMA's Decision No 15/2014).
- The BDL specifies that agency and Mudarabah contracts in which the bank does not have significant influence over the investments must be registered by the bank as off-balance sheet items.
- These efforts aim to strengthen the public's trust vis-à-vis Islamic banks and the year 2017 reforms focused on the losses incurred by investments financed by customer accounts pursuant to Mudarabah transac-

tions and repealed the reserve provisions of BDL's Basic Decision No 8828/2004.

- These legal frameworks should contribute to improve trust in Islamic banks. Also noteworthy is that the recent BDL decisions have focused on Mudarabah, an equity-based contract that is at the heart of Islamic finance. However, despite these reforms, Islamic finance still faces a number of challenges in Lebanon such as the absence of appropriate tax schemes, which the parliament only can introduce, as well as the lack of proper culture and education of the public by the banks on Islamic products.

Global Strategy for Islamic Finance

The World Bank Group approach to Islamic finance is associated with Banks in order to reduce poverty, improve finance sector and to build financial sector's stability (and resilience) in Entities. By helping to expand the use of Sharia-compliant Islamic financial activities. The World Bank Group helps to deliver benefits to Entities in the following strategic areas:

- Sustainable development of Islamic finance: opens the possibilities for economic growth, eradicating poverty and sharing prosperity. This approach contributes to economic development, giving it access to assets and the real economy. The use of profit/loss-sharing transactions' approach, encourages the provision of capital to productive *Entities* that can increase over-all benefits. Here, the emphasis is on tangible assets, which ensures that *Entities* support transactions that serve a real/concrete purpose, thus blocking and/or locating SOGFP speculations.
- Helps to promote Entities' financial sector development, by expanding the range and reach of financial services/products. It also improves financial access of deprived people to financial services. It can improve agricultural finance, by to improving food security. Knowing that of 1.6 billion Moslems in the world, only 14% of them, use any type of banking systems.
- It strengthens financial stability, like in the case of the 2008 global financial crisis, which ravaged most of the world's financial systems, whereas Islamic financial *Entities* were relatively untouched even protected by their fundamental principles of GGRisk-sharing and the

avoidance of leverage and blocking SOGFP speculative products, services and actions.

IFSVLC's Critical Success Factors

Figure 17. IFSVLC's CSFs that have an average of 8.25.

Critical Success Factors	KPIs	Weightings
CSF_IFSVLC_FinTech_Constraints	Possible	From 1 to 10. 09 Selected
CSF_IFSVLC_Structure_Evolution	Possible	From 1 to 10. 09 Selected
CSF_IFSVLC_Laws_Governance	Complex	From 1 to 10. 08 Selected
CSF_IFSVLC_Influence_LFS	VeryComplex	From 1 to 10. 07 Selected

As shown in Figure 17, the result is 8.25 (Risky).

THE PROTOTYPE'S INTEGRATION

The ACS4GG Background

As already presented in the ACS4GG (or Figure 3) are cases related to Islamic-finance and its Entities' status in the MEA and the possibility to integrate in mixed-environments, like Lebanon and Byrut. Byrut has an exposed cultural and ethnical mixture that made it very attractive and promoted it as a leading Financial-hub. Byrut's financial-predispositions attracted many institutions (financial and other…) and personalities to create and promote their businesses (Trad, 2018e). At the same time, it is unstable and depends on:

- Byrut's cosmopolitan features based on a unique ethnically diverse culture, economy and LFS is supported by a constitution.
- A constitution with a predominant liberal and westernized banking system that was designed by Michel Sursock and other prominent Lebanese personalities like Petro Trad.
- Its specific diverse and cosmopolitan characteristics, might provoke some its Pan-Arab neighbors.

- Carefully Integrate FinTech and block SOGFP activities.
- MEA's gas and oil fields brought wealth, and many other GGProblems related to Heavens' envies.
- Unifying information, using secured networks.
- Lebanon has an ultra-liberal economical model that contradicts Project's main vision and goals which are mainly not commercial objectives.
- To be able to conduct a Project for the extremely complex region of Byrut, that is a part of the Mont-Lebanon's region.
- To deliver a robust strategy against mainly SOGFP fatal blows.
- The idea of a financial Franco Lebanese Entity melted down in front of a ruthless ethno-religious war; financed by the peace-loving SOGFP oriented Heavens.
- The Lebanese (un)civil war (1975 to 1991), destroyed its infrastructure and LFS.
- This suspiciously random GGEvents catapulted other Heavens, which a classical SOGFP behaviour.
- The looted values and assets were transferred to SOGFP banks…

The Proof of Concept

The PoC presents the IFSVLC mechanics that interface the DMS4GG, which uses the internal initial Project (or phase 1) sets of CSFs' which were processed and evaluated in the figures. IFSVLC concept defines relationships between the Projects requirements and Entity's CSAs/CSFs, which use Global Unique Identifiers (GUID). The PoC is implemented by using the IHIPTF client's interface that is shown in Figure 18; where the starting activity is to structure the organizational part for an Entity.

Figure 18. The IHIPTF's client interaction.

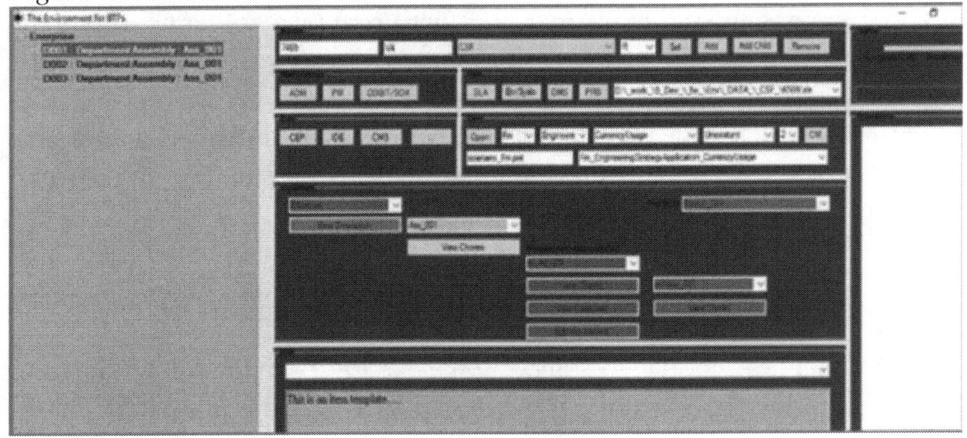

After starting IHIPTF*'s* graphical interface, the sets of CSFs are selected. Then follows CSFs' attachment to a specific node of the IHIPTF*'s* graphical HDT; to link later the microartefacts. These scripts make up the intelligence basis and the AHMM4GG's instance set of actions that are processed in the background to support the GGC. AHMM4GG's main constraint, to implement the IFSVLC is that CSA_DTs having an average result below 8.5 will be ignored. As shown in Figure 19, the overall IFSVLC research's average is 7,7 and drops the ones in red. Which means that IFSVLC's integration is "Complex".

Figure 19. IFSVLC research's outcome is 7,71.

CSA Category of CSFs/KPIs	Transformation Capability	Average Result	Table
The RDP's Integration	Complex	From 1 to 10. 8.0	[illegible]
The ACS4GGs Usage	Complex	From 1 to 10. 7.25	[illegible]
The Role of AHMM4GG	Complex	From 1 to 10. 8.0	[illegible]
The Role of ICS/FinTech	VeryComplex	From 1 to 10. 7.0	[illegible]
The Role of AI and Intelligence	Complex	From 1 to 10. 7.75	[illegible]
The Role of IFSVLC	Complex	From 1 to 10. 8.25	6
Phase's 1 Outcome	Complex	From 1 to 10. 7.71	[illegible]
Evaluate First Phase			

Linking ACSs: Integration and Unification

Phase 1, showed that the Project is "Complex" and Phase 2 starts with the selection of the ACS4GGs from the following list of cases:

- Finance, governance, law and technology category or the ACS4GG_FinTech, ACS4GG_DemoIm, ACS4GG_IdeoPol and ACS4GG_IoT were selected.
- SOGFP crimes that were coordinated by corrupt Lebanese personalities like Riad Salamé, Beidas, and Heavens (Swiss banks), plundered and destroyed the LFS Lebanon (Courrier International, 2024 ; OLJ, 2023, 2024).
- The ArchiSurance that has an archaic ICS, a mainframe, claim files service, customer file service. The ACS4GG manages claims activities where the basic application is used for the PoC's implementation (Trad, & Kalpić, 2018c).
- And other as presented in the ACS4GG CSA.

The IFSVLC defines relationships between the GGC, requirements, CSFs, and FinTech-services.

Experiment's Processing on a Concrete Tree Node

In phase 2, the HDT is used, to find a combination of HDT's action, used to solve a GGProblem related to this chapter's RQ. A specifically selected CSF is linked to a GGProblem type and a related set of actions; where the processing starts in the root node. Each GGProblem, like this case the PRB_IFSVLC_SOGFP GGProblem, has the following set of actions:

- ACT_IFSVLC_Define_SOGFP_ProblemType
- ACT_IFSVLC_Verify_SOGFP_ProblemType
- ACT_IFSVLC_Match_SOGFP_ProblemType
- ACT_IFSVLC_Validate_SOGFP_ProblemType

For this PoC, the CSF_IFSVLC_SOGFP was selected as an active CSF and the goal is to find solutions for related GGProblem(s). GGProblems are processed by the HDT and that involves the definition of a set of actions and solutions. The HDT was applied to solve the CSF_IFSVLC_SOGFP related GGProblem (PRB_IFSVLC_SOGFP), which is solved by the following steps:

- Setup the RDP and initialize its resources.
- Finalize LRP and select relevant resources and GGItems.
- The AHMM4GG offer sets of predefined and its related CSAs/CSFs.
- Map CSF to sets of actions and constraints.
- The NLP4GG-scripts make up the processing logic of the IFSVLC defined GGProblems.
- Sets of actions are processed in the background by the use of FinTech-services.
- Use EA to model the selected CSA4GGs.
- Relating ACS4GGs' CSFs, and transactions' to CSF_IFSVLC_SOGFP (done in phase 1).
- The HDT engine is configured, weighted and tuned using configuration information.
- Linking HDT-node to the pseudo-quantitative module and deliver the initial state that is the root-node.
- The HDT starts with the initial CSF_IFSVLC_SOGFP and offers a set of solutions.
- Then CSFs' are attached to specific HDT-nodes and sets of actions.
- The HDT uses the input from previous phases to propose optimal solutions.
- The selected GGProblem is related to *the detection of SOGFP activities*.
- If new GGProblem(s) are encountered then the same process is executed.
- This is a set of recursive actions until a solution is found.

SOLUTION AND RECOMMENDATIONS

This chapter's main goal is to present the IFSVLC and the GGC to support the MEA in general and more specifically Lebanon and its LFS. The ability of a traditional Entity to transform into a dynamic FinTech-services, ensures long-term financial-sustainability. Because of the low score, Figure 19, shows that IFSVLC's implementation "Complex", and the recommendations are:

- The Manager must have cross-functional skills and a strong GGC/GCC capacities.
- Unbundle Entity's services to deliver the needed FinTech-services.
- ADM's integration with the IFSVLC enables the automation of its interfaces.
- Select IFSVLC's CSFs, KPIs and CSAs; and weighting/rating rules.
- An IFSVLC must be established and used to check Entity's status.
- Define the interface to interact Intelligence.
- A restructured Entity's Financial-hub will easily integrate in the global economy.
- Enforce IHI solutions by promoting hands-on skills.
- Islamic-finance (and banking) have progressed, especially in the pre-dominately in Moslem Entities.
- Islamic-finance and banking, face strong resistances in mixed Entities.

CONCLUSION

The PoC used CSAs/CSFs links to specific Project resources and the HDT evaluated these CSFs. The result implies showed that IFSVLC based transformation is "Complex" and needs strong leaders. This chapter is part of a series of works related to Projects and SOGFPs; and it focuses on transformational activities. In this chapter the focus is on IFSVLC and GGC, which proposes a strategy to avoid financial-crimes and locked-in situations. Such activities extract trillions by Heavens, cause conflicts and various looting schemes. The Nobel prize winner, the British economist, Angus Deatoon, warns about the destructive SOGFPs professionals graduating from business schools, who cause major damages. Deatoon recommends stopping this type of financial brutalities (Le Monde, 2019). Actual international laws cannot prevent such an attitude, which has immense negative impacts (Clarke & Tigue, 1975); which can imply that the world governing organism are corrupt and participating in these global crimes. The evolution of ethics in finance might bring an end to such financial manipulations and eventually bring to trial Entities (Heavens), for committing major crimes against humanity. These crimes caused the deaths of hundreds of thousands of people and the looting of their goods, like in WWII, Lebanon... Knowing that countries

REFERENCES

Agger, I., & Jensen, S. (1996). *Trauma and Healing Under State Terrorism*. ZEB Books.

Agievich, V. (2014). Mathematical model and multi-criteria analysis of designing large-scale enterprise roadmap. PhD thesis on the specialty 05.13.18 – Mathematical modelling, numerical methods and complexes of programs.

Albawaba Business. (2014). Why Islamic Finance is failing in Lebanon. Albawaba Business. https://www.albawaba.com/business/lebanon-islamic-finance-545264

Alderman, L. (2019). French Court Fines UBS $4.2 Billion for Helping Clients Evade Taxes. The New York Times. USA. Retrieved from https://www.nytimes.com/2019/02/20/business/ubs -france-tax-evasion.html

Bruce, C. (1994). Supervising LRPs. UK. In Zuber-Skerritt, O., & Ryan, Y. (Eds.), *Quality in postgraduate education*. Kogan Page.

BSI. (2015). *Architectural framework for the Internet of Things, for Smart Cities*. BSI.

Burton, B., & Burke, B. (2012). *EA in the Arab Gulf States: Increased Focus on Delivering Strategic Value. Published: 3 August 2012*. Gartner Inc.

Chammas, G. (2006). Islamic Finance Industry In Lebanon: Horizons, Enhancements And Projections. A thesis. Ecole Supérieure des Affaires (ESA)-Lebanon.

Chedid, E. (2018). Regulatory updates on Islamic banking in Lebanon. Dentons. Lebanon. https://www.dentons.com/en/insights/alerts/2018/august/15/regulatory-updates-on-islamic-banking-in-lebanon

Clarke, Th., & Tigue, J. (1975). *Dirty money: Swiss banks, the Mafia, money laundering, and white-collar crime*. Simon and Schuster.

Courrier International. (2024). Crise. Le "Madoff libanais" Riad Salamé enfin derrière les barreaux: un "tour de passe-passe"? Courrier International. https://www.courrierinternational.com/article/crise-le-madoff-libanais-riad-salame-enfin-derriere-les-barreaux-un-tour-de-passe-passe_221826

Farhoomand, A., & Lentini, D. (2008). e-Business Transformation in the Banking Industry: The Case of Citibank. Asia Case Research Centre. The University of Hong Kong. Retrieved from http://www.acrc.hku.hk/case/case_showdetails.asp?ct=newly&c=944&cp=1949&pt=1

Fisk, R. (2011). Robert Fisk: Phoenicians footprints all over Beirut. https://www.independent.co.uk/voices/commentators/fisk/robert-fisk-phoenician-footprints-all-over-beirut-6271510.html. Independent.

Gabrys, E., & Field, S. Gartner, (2014). CIO Agenda: A Gulf Cooperation Council Perspective. Published: 20 March 2014. Gartner Inc. USA.

Geranmayeh, E. (2018). *Regional Geopolitical Rivalries in the Middle East: Implications for Europe. IAI-FEPS. Istituto Affari Internazionali (IAI) and Foundation for European Progressive Studies*. FEPS.

Gomart, Th. (2016). *The Return Of Geopolitical Risk-Russia, China and the United States. Institut français des relations internationales*. Ifri.

Gunasekare, U. (2015). Mixed Research Method as the Third Research Paradigm: A Literature Review. Volume 4 Issue 8, August 2015. University of Kelaniya. IJSR.

Hamadou, I., Yumna, A., Hamadou, H., & Jallow, M. (2024). Unleashing the power of artificial intelligence in Islamic banking: A case study of Bank Syariah Indonesia (BSI). MF-Journal. https://mf-journal.com/article/view/116

Hussain, M., Nadeem, M. W., Iqbal, S., Mehrban, S., Fatima, S. N., Hakeem, O., & Mustafa, G. (2019). Security and Privacy in FinTech: A Policy Enforcement Framework. In Rafay, A. (Ed.), *FinTech as a Disruptive Technology for Financial Institutions* (pp. 81–97). IGI Global. DOI: 10.4018/978-1-5225-7805-5.ch005

Jonkers, H., Band, I., & Quartel, D. (2012a). *ArchiSurance Case Study*. The Open Group.

Joseph, Ch. (2014). Types of eCommerce Business Models. https://smallbusiness.chron.com/types-ecommerce-business-models-2447.html. Demand Media.

Kagan, J. (2020). Financial Technology – FinTech. Investopedia. https://www.investopedia.com/terms/f/FinTech.asp

Kassir, K. (2010). *Beirut*. University of California Press.

Keraine, R. (2019). The Fundamentals of Islamic Finance. INVIVOO. https://blog.invivoo.com/the-fundamentals-of-islamic-finance/

Kismawadi, E., Irfan, M., Abdul, S., & Shah, R. (2023). *Revolutionizing Islamic Finance: Artificial Intelligence's Role in the Future of Industry. Book: The Impact of AI Innovation on Financial Sectors in the Era of Industry 5.0 (/book/impact-innovation-financial-sectors-era/321132)*. IGI-Global., DOI: 10.4018/979-8-3693-0082-4.ch011

Kismawadi, R., Aditchere, J., & Libeesh, P. (2024). Integration of Artificial Intelligence Technology in Islamic Financial Risk Management for Sustainable Development. Applications of Block Chain technology and Artificial Intelligence. pp 53–71. https://link.springer.com/chapter/10.1007/978-3-031-47324-1_4

Kolbe, Th. (2015). Smart Models for Smart Cities - Modeling of Dynamics, Sensors, Urban Indicators, and Planning Actions. 29th of October 2015 Joint International Geoinformation Conference JIGC 2015, Kuala Lumpur.

Kowall, J., & Fletcher, C. (2013). *Modernize Your Monitoring Strategy by Combining Unified Monitoring and Log Analytics Tools*. Gartner Inc.

KPMG. (2020). Top 10 FinTech predictions for 2020. KPMG. https://home.kpmg/xx/en/home/campaigns/2020/02/pulse-of-FinTech-h2-19-top-10-predictions-for-2020.html

Lazar, I., Motogna, S., & Parv, B. (2010). Behaviour-Driven Development of Foundational UML Components. Department of Computer Science. Babes-Bolyai University. Cluj-Napoca, Romania. DOI: 10.1016/j.entcs.2010.07.007

Le Monde. (2019). Le Prix Nobel d'économie Angus Deaton: Quand l'Etat produit une élite prédatrice [Nobel Lauréate in Economics Angus Deaton: "When the state produces a predatory elite]. Le Monde. Retrieved from https://www.lemonde.fr/idees/article/2019/12/27/angus-deaton -quand-l-etat-produit-une-elite-predatrice_6024205_3232.html

Lea, R. (2017). Smart City Standards: An overview.

Makarchenko, M., Nerkararian, S., & Shmeleva, S. (2016). How Traditional Banks Should Work in Smart City. Communications in Computer and Information Science. DOI: . Conference: International Conference on Digital Transformation and Global Society.DOI: 10.1007/978-3-319-49700-6_13

Malhotra, E. (2005). Integrating knowledge management technologies in organizational business processes: getting real time enterprises to deliver real business performance. Emerald Group Publishing Limited, Journal of Knowledge Management. USA.

Markides, C. (2011, March). Crossing the Chasm: How to Convert Relevant Research Into Managerially Useful Research. [London, UK.]. *The Journal of Applied Behavioral Science*, *47*(1), 121–134. DOI: 10.1177/0021886310388162

MENA-OECD. (2018). *Background Note: Country case studies: Building economic resilience in Lebanon and Libya. Resilience in fragile situations. Mena-oecd economic/resilience task force.* Islamic Development Bank.

Miori, V., & Russo, D. (2014). Domotic Evolution towards the IoT. IEEE. *28th International Conference on Advanced Information Networking and Applications Workshops.* DOI: . Victoria, BC, Canada.DOI: 10.1109/WAINA.2014.128

Morrison, M. (2016). Critical Success Factors – Analysis made easy, a step by step guide. rapidBI.

Nakamoto, S. (2008). Bitcoin: A Peer-to-Peer Electronic Cash System. www.bitcoin.org. BITCOIN.

OASIS. (2014). *ISO/IEC and OASIS Collaborate on E-Business Standards-Standards Groups Increase Cross-Participation to Enhance Interoperability.* The OASIS Group.

OLJ. (2023). Des centaines de millions de dollars que Riad Salamé est accusé d'avoir détournés auraient atterri en Suisse. https://www.lorientlejour.com/article/1329590/les-fonds-detournees-de-la-banque-centrale-du-liban-ont-ete-transferes-en-suisse-media.html

OLJ. (2024). Crimes Financiers Au Liban-Mandat d'arrêt contre Riad Salamé, qui reste en détention à l'issue de son interrogatoire/La cheffe du contentieux de l'Etat, partie civile dans l'affaire, a été empêchée d'assister à l'audience. https://www.lorientlejour.com/article/1426489/riad-salame-est-arrive-au-palais-de-justice-sous-les-huees-des-manifestants.html?utm_source=olj&%E2%80%A6

Penn, A., & Arias, M. (2009). *Global E-Business Law & Taxation*. Oxford University Press. DOI: 10.1093/oso/9780195367218.001.0001

Ross, M. (2020). Working with Islamic Finance. investopedia. https://www.investopedia.com/articles/07/islamic_investing.asp

Srour, I., & Chaaban, J. (2017). *Market Research Study on the Development of Viable Economic Subsectors in Lebanon*. EU Project.

Stempel, J. (2019). UBS must defend against U.S. lawsuit over 'catastrophic' mortgage losses. Yahoo Finance. Retrieved from https://finance.yahoo.com/news/ubs-must-defend-against-u -214743943.html

Stupples, B., Sazonov, A., & Woolley, S. (2019, July 26). UBS Whistle-Blower Hunts Trillions Hidden in Treasure Isles. Bloomberg. Retrieved from https://www.bloomberg.com/news/articles/2019-07-26/ubs-whistle-blower-hunts-trillions-hidden-in-treasure-islands

Sujud, H., & Hachem, B. (2018). Reality and future of islamic banking in lebanon. *European Journal of Scientific Research*.

The Open Group. (2011a). *The Open Group's Architecture Framework*.

The Open Group. (2011b). *Architecture Development Method*. The Open Group.

Trad, A. (2018e). *The Business Transformation and Enterprise Architecture Framework Applied to analyse-The historically recent Rise and the 1975 Fall of the Lebanese Business Ecosystem*. IGI-Global.

Trad, A., & Kalpić, D. (2017b). *A Neural Networks Portable and Agnostic Implementation IHIPTF for Business Transformation Projects. The Basic Structure*. IEEE.

Trad, A., & Kalpić, D. (2017c). *A Neural Networks Portable and Agnostic Implementation IHIPTF for Business Transformation Projects. The Framework*. IEEE.

Trad, A., & Kalpić, D. (2018a). *The Business Transformation Framework and Enterprise Architecture Framework for Managers in Business Innovation-Knowledge Management in Global Software Engineering (HKMS)*. IGI-Global.

Trad, A., & Kalpić, D. (2018c). The Business Transformation Framework and Enterprise Architecture Framework for Managers in Business Innovation. The role of legacy processes in automated business environments. The Proceedings of E-LEADER 2017 Berlin, 1.

Trad, A., & Kalpić, D. (2018f). *An applied mathematical model for business transformation-The Holistic Critical Success Factors Management System (HCSFMS). Encyclopaedia of E-Commerce Development, Implementation, and Management*. IGI-Global.

Trad, A., & Kalpić, D. (2019a). The Business Transformation Framework and the Application of a Holistic Strategic Security Concept. Chinese American Scholars Association Conference E-Leader, Conference, Brno. Check Republic.

Trad, A., & Kalpić, D. (2020a). *Using Applied Mathematical Models for Business Transformation. Author Book*. IGI-Global. DOI: 10.4018/978-1-7998-1009-4

Trad, A., Nakitende, M., & Oke, T. (2021). *Tech-Based Enterprise Control and Audit for Financial Crimes: The Case of State-Owned Global Financial Predators (SOGFP). Book: Handbook of Research on Theory and Practice of Financial Crimes*. IGI Global.

UN. (2020). Lebanon: UN-backed tribunal sentences Hezbollah militant in Hariri assassination. UN. https://news.un.org/en/story/2020/12/1079892

UN. (2023). Justice served: Lebanon's Special Tribunal closes. UN. https://news.un.org/en/story/2023/12/1145217

Wikipedia. (2020a). Islamic banking and finance. Wikipedia, the free encyclopedia. https://en.wikipedia.org/wiki/Islamic_banking_and_finance

Wikipedia. (2024a). Raymond Eddé. https://en.wikipedia.org/wiki/Raymond_Edd%C3%A9

Zalloua, P. (2004). The NGM Study "Who were the SPs" and the Return of the SPs. Interview Lebanese Broadcasting Corporation (LBC). Retrieved April 16, 2017, from. Phoenicia. National Geographic (October 2004).

Chapter 4
Optimizing Operational Efficiency in Islamic Banking Integrating Data Envelopment Analysis and Monte Carlo Simulations:
Optimizing Operational Efficiency in Islamic Banking

Muhammet Enis Bulak
https://orcid.org/0000-0003-3784-7830
Uskudar University, Turkey

Mohammad Ali Chebli
Uskudar University, Turkey

ABSTRACT

Operational efficiency is crucial for the competitiveness and sustainability of banks, particularly in Islamic banking, which adheres to principles such as the prohibition of interest (riba) and asset-backed financing. This study integrates Data Envelopment Analysis (DEA) with Monte Carlo (MC) simulations to evaluate the operational efficiency of Islamic banks. It focuses

DOI: 10.4018/979-8-3693-8079-6.ch004

on three key analyses: assessing Tamweel Bank's efficiency from 2013 to 2023 using historical financial data; generating simulated data with MC simulations to stabilize efficiency estimates; and comparing Tamweel and Alizz Islamic Banks for 2023. Both Constant Returns to Scale (CCR) and Variable Returns to Scale (BCC) DEA models are applied to capture diverse efficiency dimensions. Findings show significant variability in Tamweel Bank's efficiency, peaking in 2019 but declining by 2023. MC simulations provide an aggregated efficiency benchmark, emphasizing the value of combining deterministic and stochastic methods. The comparative analysis highlights an efficiency gap, offering strategic insights for improvement.

INTRODUCTION

Operational efficiency stands as a cornerstone of competitiveness and sustainability within the banking sector. For financial institutions, the ability to maximize outputs while minimizing inputs not only enhances profitability but also ensures resilience in fluctuating economic landscapes. In the realm of Islamic banking, which operates under distinct principles such as the prohibition of interest (riba), avoidance of speculative activities (gharar), and the requirement for asset-backed financing, assessing operational efficiency presents unique challenges and opportunities. These principles necessitate specialized frameworks for efficiency evaluation, differentiating Islamic banks from their conventional counterparts.

This study seeks to provide a comprehensive evaluation of operational efficiency in Islamic banks by employing a dual-methodological approach that integrates Data Envelopment Analysis (DEA) with Monte Carlo (MC) simulation. DEA is a non-parametric technique renowned for its ability to benchmark the relative efficiency of decision-making units (DMUs) based on multiple inputs and outputs without requiring predefined weights. However, DEA's deterministic nature can limit its capacity to account for the inherent uncertainties and variability present in financial data. To address this limitation, Monte Carlo simulations are incorporated to generate synthetic datasets that introduce stochastic elements, thereby enhancing the robustness and generalizability of the efficiency assessments.

The methodology is systematically structured into three primary analyses:

1. Efficiency Analysis Using Historical Financial Data (2013–2023): This analysis employs DEA to assess the operational efficiency of selected Islamic banks over a ten-year period. By examining temporal trends, the study identifies periods of peak performance and phases of declining efficiency, providing insights into the dynamic nature of operational practices and external economic influences.
2. Efficiency Analysis Using Simulated Data Generated through Monte Carlo Simulation (2013–2023): Leveraging Monte Carlo simulations, this analysis generates synthetic datasets that account for variability and uncertainty in financial metrics. By integrating these simulations with DEA, the study produces a stabilized, long-term efficiency estimate that complements the year-specific insights derived from historical data.
3. Comparative Efficiency Analysis of Multiple Islamic Banks for 2023: This comparative analysis benchmarks the efficiency of different Islamic banks within a single year, identifying performance gaps and best practices. By contrasting the efficiency scores of various institutions, the study highlights areas for strategic improvement and operational optimization.

The integration of deterministic and stochastic methods, coupled with robust DEA modeling under both Constant Returns to Scale (CCR) and Variable Returns to Scale (BCC), ensures a holistic evaluation of operational efficiency. This multifaceted framework not only addresses historical performance but also incorporates simulated uncertainties, offering a comprehensive understanding of efficiency dynamics within Islamic banking.

The significance of this study lies in its methodological innovation and practical implications. By combining DEA with Monte Carlo simulations, the research introduces a more nuanced and resilient approach to efficiency evaluation, adaptable to the unique operational frameworks of Islamic banks. The findings provide actionable insights for bank managers and policymakers aiming to optimize resource utilization, enhance profitability, and maintain a competitive edge in the evolving financial landscape.

In the subsequent sections, the methodology is detailed, followed by the presentation of results from the three analytical models. The study concludes with a discussion of key findings, implications for Islamic banking, and recommendations for future research, thereby contributing to the broader discourse on financial efficiency evaluation in specialized banking sectors.

LITERATURE REVIEW

Operational efficiency is a cornerstone of banking competitiveness, enabling institutions to optimize their outputs, such as profits and market share, while minimizing inputs like labor and operational costs. Banks that achieve high operational efficiency are better positioned to withstand economic volatility and maintain their roles as financial intermediaries (Nguyen & Tran, 2021; Hussain & Imran, 2020). Key areas influencing operational efficiency include cost management, revenue diversification, and asset utilization, all of which are essential for navigating the dynamic regulatory and market landscape of modern banking (Ali & Abbas, 2022; Zhang et al., 2022). Efficient banks consistently outperform their peers, demonstrating superior profitability and resilience (Chen et al., 2021; Gao et al., 2023).

Data Envelopment Analysis (DEA) has emerged as a widely adopted methodology for assessing efficiency in banking. Its non-parametric nature allows for the evaluation of banks by benchmarking their performance against a theoretical "efficiency frontier" without relying on predefined weights for inputs and outputs (Charnes et al., 1978; Liu & Huang, 2021). DEA has been instrumental in providing insights into operational efficiency, highlighting opportunities for cost reduction and revenue enhancement (Ahmad et al., 2021; Chen et al., 2021). Recent methodological advancements, such as network DEA models, have enhanced its applicability by breaking down banking operations into distinct sub-processes, enabling a granular analysis of efficiency at different operational stages (Chen et al., 2021). Additionally, integrating DEA with stochastic approaches like Monte Carlo simulations addresses some of its limitations by accounting for data variability, leading to more robust and reliable results (Singh & Kaur, 2022; Gao et al., 2023).

Monte Carlo (MC) simulations complement DEA by modeling uncertainty and variability in banking data. Through the generation of synthetic datasets, MC simulations provide a framework for stress testing, allowing banks to evaluate the potential impact of adverse conditions on their performance (Rahman & Ali, 2021; Zhou & Wu, 2022). MC simulations have also been applied for performance forecasting, aiding in strategic planning and resource allocation (Liu & Huang, 2021). The integration of DEA with MC simulations enhances the robustness of efficiency assessments by incorporating stochastic elements into the analysis, which is especially useful in addressing DEA's sensitivity to outliers and deterministic assumptions (Gao et al., 2023;

Hussain & Imran, 2020). For example, hybrid DEA-Monte Carlo models and bootstrap DEA techniques have introduced statistical confidence intervals around efficiency scores, providing a more comprehensive evaluation framework (Patel & Kumar, 2022).

Islamic banking, guided by Shariah principles, presents unique challenges and opportunities for efficiency assessment. Unlike conventional banking, Islamic banking prohibits interest-based transactions and emphasizes risk-sharing, which necessitates specialized frameworks for evaluating operational efficiency (Iqbal & Molyneux, 2021; El-Gamal, 2020). DEA has proven effective in evaluating the efficiency of Islamic banks by focusing on dimensions such as profitability, liquidity, and adherence to Shariah compliance (Ahmad et al., 2021; Mohammad et al., 2023). However, comparative studies have revealed significant variations in efficiency levels between Islamic and conventional banks, often driven by differences in operational models and regulatory frameworks (Ali & Abbas, 2022; Khan & Rahman, 2023). Despite its strengths, efficiency assessment in Islamic banking is hindered by limited availability of standardized data and the need to adapt traditional DEA methodologies to Shariah-compliant frameworks (Singh & Kaur, 2022; Mirakhor & Iqbal, 2021).

The integration of DEA with Monte Carlo simulations has been recognized as a significant methodological advancement for efficiency assessments in banking. By combining DEA's ability to benchmark performance with MC simulations' capacity to model data uncertainty, this hybrid approach provides robust efficiency metrics (Gao et al., 2023). Recent applications of this integration include evaluating Islamic banks' efficiency in the context of stochastic variability and developing more resilient performance models (Ahmad et al., 2021; Wang et al., 2021). The hybrid DEA-MC approach has also proven effective in other industries, such as healthcare and manufacturing, demonstrating its versatility (Hassan et al., 2023).

Benchmarking is another critical aspect of banking efficiency, allowing institutions to compare their performance against industry standards and peers. Commonly used metrics include cost-to-income ratios, return on assets (ROA), and return on equity (ROE) (Nguyen & Tran, 2021; Zhang et al., 2022). DEA is particularly well-suited for benchmarking due to its ability to rank institutions based on relative efficiency without requiring subjective input weights (Liu & Huang, 2021; Ahmad et al., 2021). Recent developments in benchmarking methodologies include the integration of sustainability metrics, reflecting the growing emphasis on environmental and social responsibility in banking (Mohammad et al., 2023). Dynamic benchmarking models, enabled by advancements in DEA, facilitate real-time performance monitoring and allow for timely strategic adjustments (Chen et al., 2021).

In conclusion, operational efficiency remains a vital determinant of banking performance, with DEA serving as a pivotal tool for its assessment. The integration of DEA with Monte Carlo simulations has addressed critical gaps in existing methodologies, particularly by enhancing robustness and accommodating data variability. These advancements are especially significant for Islamic banking, where unique operational principles demand specialized evaluation frameworks. As the financial landscape evolves, incorporating dynamic and stochastic methods into efficiency analysis will be instrumental in fostering resilience and competitiveness in both conventional and Islamic banking. This study contributes to the existing literature by integrating DEA and Monte Carlo simulations to evaluate efficiency comprehensively, offering actionable insights for strategic improvement and operational resilience.

METHODOLOGY

This study employs a comprehensive and rigorous methodology integrating Data Envelopment Analysis (DEA) and Monte Carlo (MC) simulation to evaluate and compare the operational efficiency of Islamic banks. The methodology is systematically structured into three primary analyses:

1. Efficiency Analysis of Tamweel Bank (2013–2023) Using Historical Financial Data
2. Efficiency Analysis of Tamweel Bank (2013–2023) Using Simulated Data Generated through Monte Carlo Simulation
3. Comparative Efficiency Analysis of Tamweel Bank and Alizz Islamic Bank for 2023

The approach seamlessly combines deterministic and stochastic methods, encompassing statistical distribution analysis, synthetic data generation, and robust DEA modeling under both Constant Returns to Scale (CCR) and Variable Returns to Scale (BCC). This multifaceted framework ensures a holistic evaluation of operational efficiency, addressing both historical performance and simulated uncertainties.

1. Data Collection

Sources

Historical Data: Annual financial reports of Tamweel Bank from 2013 to 2023.

Comparative Data: Annual financial report of Alizz Islamic Bank for 2023.

Selection Criteria

As part of this study, we adopt the DEA framework to evaluate efficiency. For illustrative purposes, we include the mathematical formulation of the input-oriented CCR model, which assumes constant returns to scale. This approach aligns with our objective to demonstrate the application of DEA concisely while avoiding unnecessary complexity by limiting the scope to a representative example. The selected model emphasizes minimizing inputs while preserving output levels, providing a basis for understanding the efficiency analysis conducted in this research.

Minimize θ

$$\sum_{J+1}^{n} \lambda_j x_{ij} \leq \theta x_{ik}$$

$$\sum_{J+1}^{n} \lambda_j y_{ij} \geq y_{ik}$$

$$\lambda_j \geq 0$$

Where

- θ: Efficiency score for the DMU under evaluation.
- xijx_{ij}xij: Input iii for DMU jjj.
- yrjy_{rj}yrj: Output rrr for DMU jjj.
- λj\lambda_jλj: Weights for the reference DMUs.

Input and Output Metrics: Selected based on DEA principles to comprehensively capture the financial and operational performance of the banks. The metrics are chosen to reflect key aspects of liquidity, investment activities, asset management, liabilities, and profitability.

Inputs:

Bank Balances and Cash (AED '000): Reflects liquidity and cash management.

Islamic Financing and Investing Assets (AED '000): Represents financing and investment activities.
Other Investments Carried at FVOCI (AED '000): Captures investment portfolio performance.
Investment Properties (AED '000): Reflects real estate investment returns.
Advances, Prepayments, and Receivables (AED '000): Indicates receivables management.
Property and Equipment (AED '000): Represents physical capital.
Accounts Payable and Other Liabilities (AED '000): Reflects obligations and liabilities.
Allocable Expenses (AED '000): Represents expenses allocated to various departments, projects, or cost centers.

Figure 1. Balancing revenue and profitability metrics

Outputs:
Gross Income (AED '000): Measures revenue generation capacity.
Profit for the Year (AED '000): Captures overall profitability.

Figure 2. Components of financial statement outputs

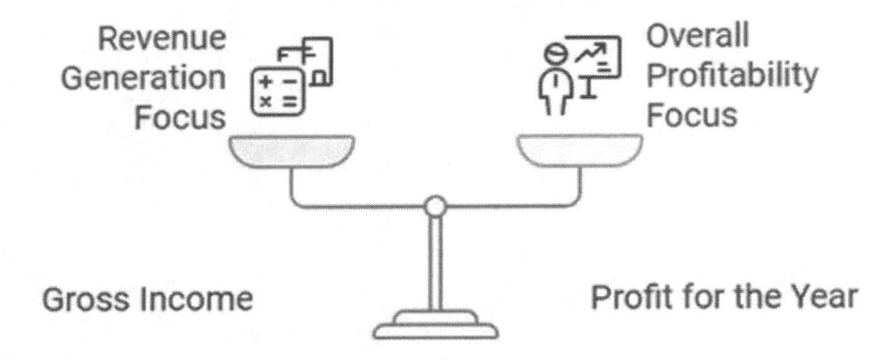

2. Statistical Distribution Analysis

To accurately model and simulate financial data, it is essential to understand the underlying probability distributions of each metric. This step ensures that the Monte Carlo simulations generate realistic and representative synthetic data.

Process:

1. Data Exploration: Each input and output metric from the historical data is examined to identify patterns, trends, and variability.
2. Goodness-of-Fit Tests: Statistical tests (e.g., Kolmogorov-Smirnov, Anderson-Darling) are conducted to determine the best-fit probability distributions for each metric. Potential distributions considered include:
    - Exponential Negative
    - Beta Binomial
    - Beta
    - Uniform Discrete
    - Gamma

3. Selection Criteria: The distribution that best fits the historical data based on the highest goodness-of-fit statistics is selected for each metric.

Outcome:

Identification of the most appropriate probability distribution for each input and output metric, forming the foundation for the Monte Carlo simulation process.

Figure 3. Refining metrics into probability distributions

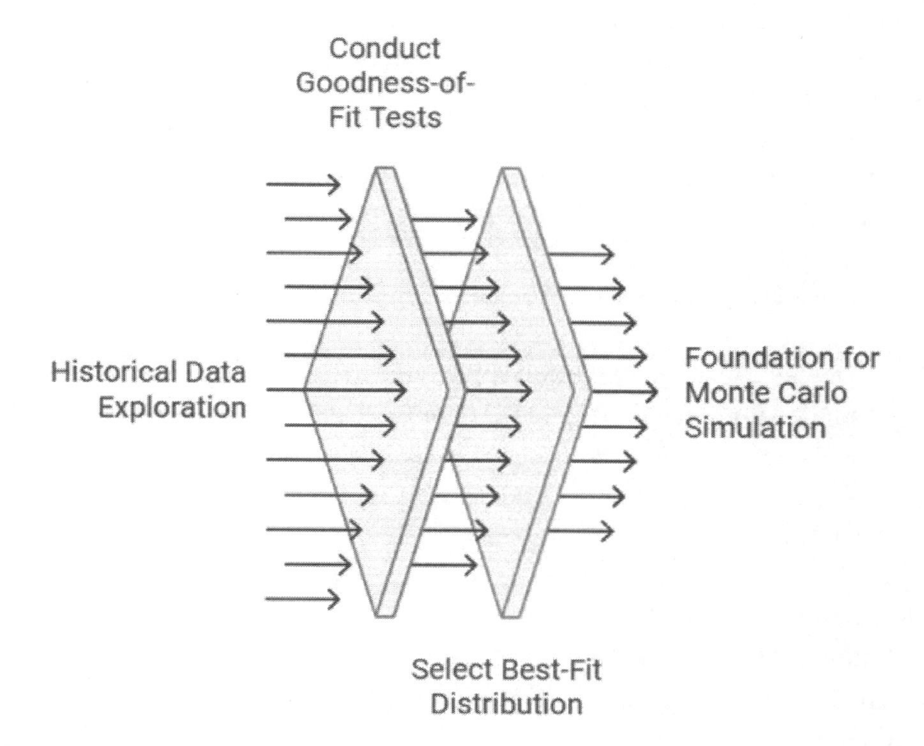

3. Monte Carlo Simulation

Monte Carlo simulations are employed to generate synthetic datasets that account for variability and uncertainty in financial metrics. This approach enhances the robustness and generalizability of the efficiency analysis by incorporating stochastic elements alongside deterministic historical data.

Simulation Steps:

1. Probability Distribution Identification: Utilize the results from the statistical distribution analysis to determine the appropriate distributions for each metric.
2. Random Sampling: Generate 10,000 random values for each input and output metric using the identified distributions. This large number of simulations ensures a comprehensive exploration of possible outcomes.
3. Synthetic Dataset Construction:

Mean Calculation: Compute the mean values of the simulated data for each metric to create a synthetic dataset representing the period 2013–2023.

Aggregation: Aggregate the synthetic data to mirror the structure and temporal span of the historical dataset, ensuring consistency in the DEA analysis.

4. Efficiency Score Generation: Apply DEA to the synthetic dataset to derive a single, aggregated efficiency score for Tamweel Bank, smoothing out year-to-year variability observed in the historical data.

Purpose:

To provide a stabilized, long-term efficiency estimate that complements the year-specific insights from historical data, facilitating strategic planning and model robustness testing.

The implementation of Data Envelopment Analysis (DEA) and Monte Carlo (MC) simulations was executed using Python, chosen for its versatility and extensive library support essential for data analysis and statistical modeling. The Pandas and NumPy libraries facilitated efficient data manipulation and numerical computations, ensuring seamless handling of both historical and synthetic datasets. Statistical distribution analyses and goodness-of-fit tests were conducted using SciPy and Statsmodels, enabling the accurate identification of appropriate probability distributions for each financial metric. For constructing and solving the DEA models under both Constant Returns to Scale (CCR) and Variable Returns to Scale (BCC) assumptions, the DEApy library was integrated with PuLP, a powerful linear programming solver. Visualization of efficiency scores and comparative analyses was achieved through Matplotlib and Seaborn, which provided clear and insightful graphical representations of the results. The entire analytical workflow was managed within Jupyter Notebook, promoting an organized and reproducible research process. Additionally, Git was employed for version control, ensuring meticulous documentation and tracking of all coding and analytical steps. 4. Software and Coding Implementation*

The implementation of Data Envelopment Analysis (DEA) and Monte Carlo (MC) simulations was executed using Python, chosen for its versatility and extensive library support essential for data analysis and statistical modeling. The Pandas and NumPy libraries facilitated efficient data manipulation and numerical computations, ensuring seamless handling of both historical and synthetic datasets. Statistical distribution analyses and goodness-of-fit tests were conducted using SciPy and Statsmodels, enabling the accurate identification of appropriate probability distributions for each financial metric. For constructing and solving the DEA models under both Constant Returns to Scale (CCR) and Variable Returns to Scale (BCC) assumptions, the DEApy library was integrated with PuLP, a powerful linear programming solver. Visualization of efficiency scores and comparative analyses was achieved through Matplotlib and Seaborn, which provided clear and insightful graphical representations of the results. The entire analytical workflow was managed within Jupyter Notebook, promoting an organized and reproducible research process. Additionally, Git was employed for version control, ensuring meticulous documentation and tracking of all coding and analytical steps.

4. Software and Coding Implementation

The implementation of Data Envelopment Analysis (DEA) and Monte Carlo (MC) simulations was executed using Python, chosen for its versatility and extensive library support essential for data analysis and statistical modeling. The Pandas and NumPy libraries facilitated efficient data manipulation and numerical computations, ensuring seamless handling of both historical and synthetic datasets. Statistical distribution analyses and goodness-of-fit tests were conducted using SciPy and Statsmodels, enabling the accurate identification of appropriate probability distributions for each financial metric. For constructing and solving the DEA models under both Constant Returns to Scale (CCR) and Variable Returns to Scale (BCC) assumptions, the DEApy library was integrated with PuLP, a powerful linear programming solver. Visualization of efficiency scores and comparative analyses was achieved through Matplotlib and Seaborn, which provided clear and insightful graphical representations of the results. The entire analytical workflow was managed within Jupyter Notebook, promoting an organized and reproducible research process. Additionally, Git was employed for version control, ensuring meticulous documentation and tracking of all coding and analytical steps.

RESULTS

Model 1: Efficiency Analysis Using Historical Data (2013–2023)

Model 1 employs Data Envelopment Analysis (DEA) to assess the operational efficiency of Tamweel Bank over a ten-year period from 2013 to 2023. The analysis differentiates between Constant Returns to Scale (CCR) and Variable Returns to Scale (BCC) efficiency scores, revealing significant variability in the bank's efficiency across different years.

CCR Efficiency Scores

2019: Achieved a perfect CCR efficiency score of 1.00, indicating optimal utilization of resources.
2023: Recorded the lowest CCR efficiency score of 0.447, highlighting substantial inefficiencies.
Trend Overview: The CCR scores fluctuated annually, with peaks in 2019 and varying efficiencies in other years, reflecting dynamic operational performance.

Figure 4. CCR output

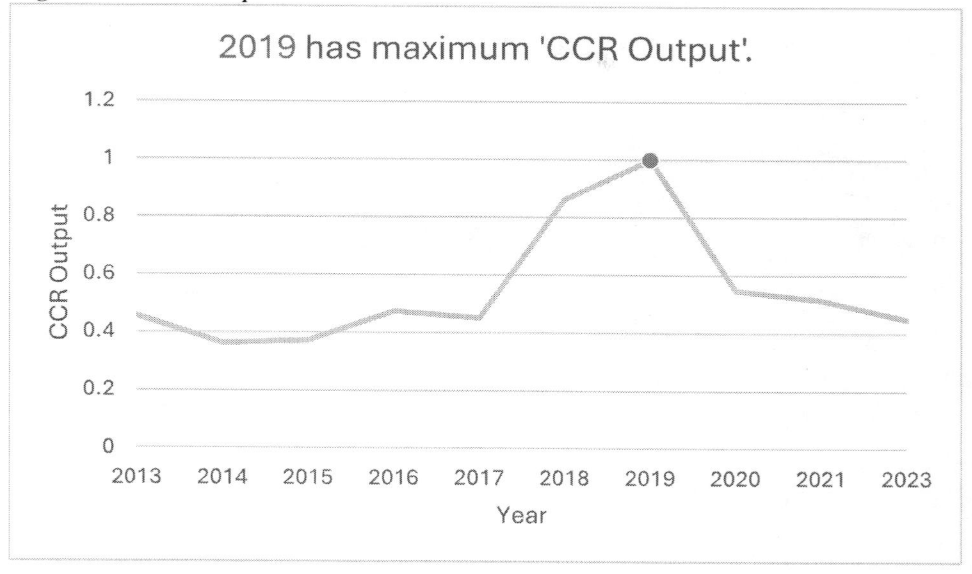

Figure 4 shows the CCR Output across the years from 2013 to 2023. It highlights that the maximum CCR Output occurred in the year 2019, reaching a peak value of 1. This is indicated by a distinct orange dot marking the highest point on the curve. The overall trend shows a gradual increase in CCR Output from 2014 to 2019, followed by a sharp decline in the subsequent years. After 2019, the output steadily decreases, stabilizing at lower levels in 2020–2023.

BCC Efficiency Scores

2019: Also attained a perfect BCC efficiency score of 1.00, underscoring optimal management under variable returns to scale.
2023: Dropped to 0.498, demonstrating inefficiencies in managing resources under changing scales.
Performance Variability: Similar to CCR, BCC scores exhibited considerable variation, peaking in 2019 and declining in subsequent years.

Figure 5. Multiple values by 'year'

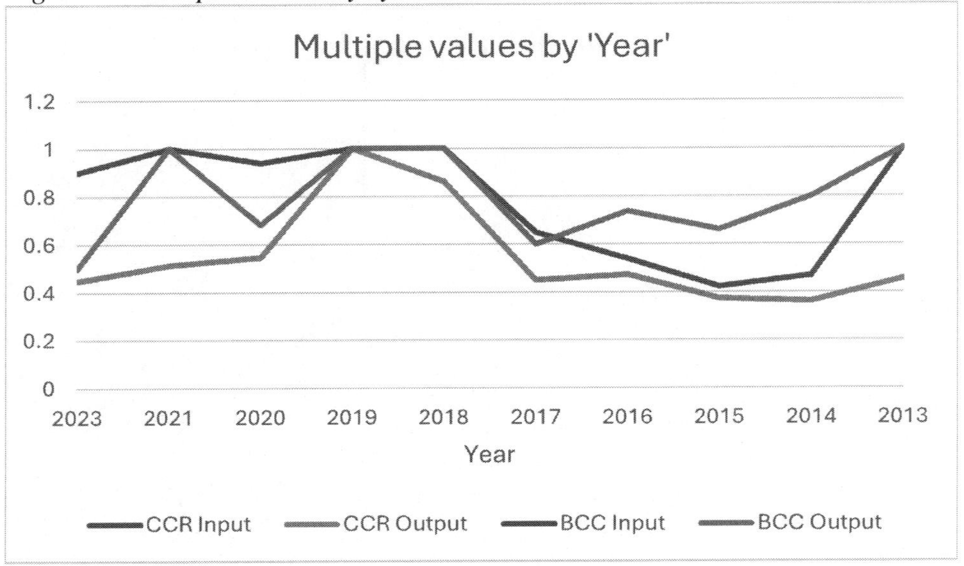

Figure 5 presents a comparison of CCR Input, CCR Output, BCC Input, and BCC Output over the years from 2013 to 2023. The BCC Input (green line) and BCC Output (purple line) generally show fluctuating trends, with peaks around 2013, 2019, and 2023. The CCR Input (blue line) and CCR Output (red line) have similar patterns, with notable peaks around 2019. After 2019, CCR Output declines sharply, whereas BCC Output and BCC Input stabilize. By 2023, BCC Output and BCC Input return to high values, indicating potential recovery or improvements. Overall, the trends reflect that the year 2019 serves as a significant turning point for multiple metrics, especially the CCR values.

Key Observations

Consistent High Performance in 2019: Both CCR and BCC scores peaked in 2019, suggesting effective resource allocation and management strategies during this year.

Declining Efficiency Post-2019: Following 2019, there was a noticeable decline in efficiency scores, reaching the lowest point in 2023. This trend may indicate emerging operational challenges or external economic factors adversely affecting performance.

Comparative Yearly Performance: For instance, in 2020, CCR and BCC scores were 0.546 and 0.682 respectively, reflecting moderate efficiency. In contrast, 2015 and 2014 exhibited lower CCR scores of 0.371 and 0.361, indicating persistent inefficiencies during these years.

This historical analysis underscores the fluctuating efficiency of Tamweel Bank, highlighting periods of both excellence and significant operational challenges.

Model 2: Efficiency Analysis Using Simulated Data

Model 2 utilizes Monte Carlo simulations to generate a synthetic dataset, providing an aggregated efficiency estimate for Tamweel Bank across the 2013–2023 period. This model offers a single, consistent efficiency score, smoothing out the year-to-year variability observed in historical data.

Simulated Efficiency Score

Value: 0.4625

Implications: This uniform efficiency score suggests that, on average, Tamweel Bank maintains a moderate level of efficiency over the decade when considering aggregated data. The simulation smoothens the fluctuations seen in historical scores, offering a generalized view of operational performance.

Input and Output Variables

Inputs:
Bank Balances and Cash: AED 316,855.67 '000
Islamic Financing and Investing Assets: AED 3,329,451.38 '000
Other Investments Carried at FVOCI: AED 54,717.68 '000
Investment Properties: AED 663,135.25 '000
Advances, Prepayments, and Receivables: AED 58,835.03 '000
Property and Equipment: AED 12,543.77 '000
Accounts Payable and Other Liabilities: AED 154,132.70 '000
Gross Income: AED 358,338.20 '000
Profit for the Year: AED 95,866.12 '000
Allocable Expenses: AED 204,428.77 '000

Figure 6. Multiple values by 'year'

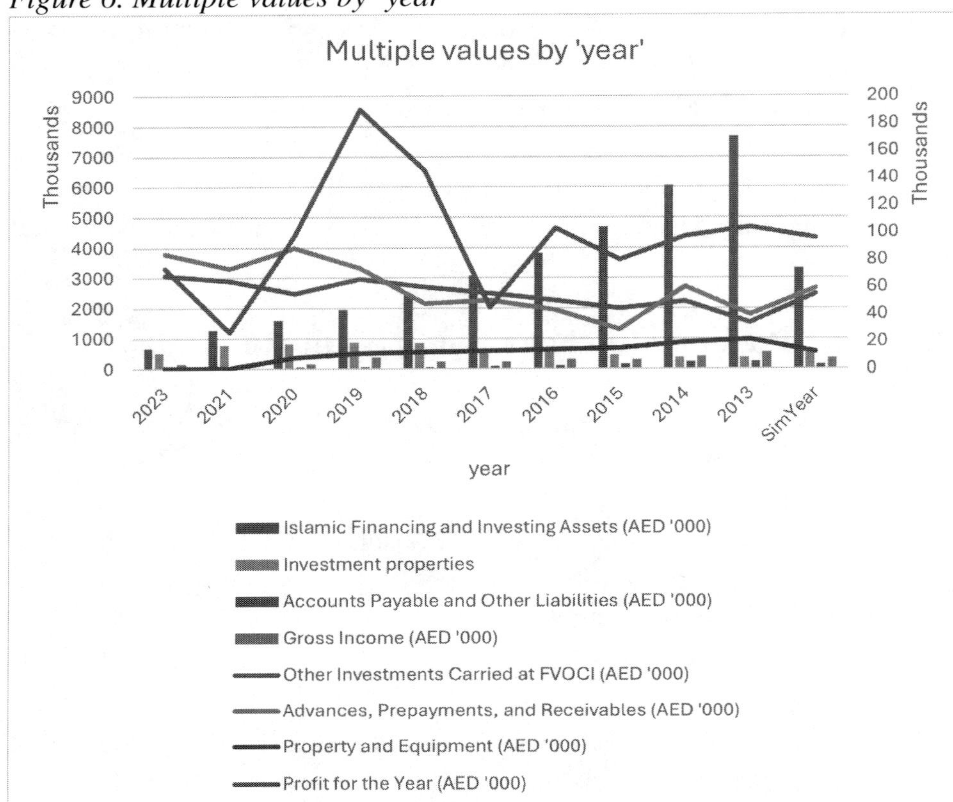

Figure 6 shows key financial metrics over time in AED '000. Profit for the Year (brown line) peaked sharply in 2018, followed by a decline and stabilization. Islamic Financing and Investing Assets (blue bars) showed growth, peaking in 2015 and 2018, before recovering again in 2023. Gross Income (orange line) remained stable but declined slightly after 2018. Other metrics, like Accounts Payable and Other Investments, stayed relatively consistent at lower levels. Overall, 2018 stands out as a peak year for financial performance.

Comparative Error Analysis

Discrepancy: The simulated efficiency score exhibits a 10% error when compared to the actual historical efficiency scores.

Impact of Stochastic Assumptions: The randomness inherent in Monte Carlo simulations contributes to this discrepancy, indicating that while the simulation provides a reliable aggregated estimate, it may not fully capture the year-specific complexities influencing actual performance.

Insights from Simulation

Long-Term Perspective: The simulated efficiency score offers a stable long-term benchmark, useful for strategic planning and evaluating overall performance trends.
Limitations: The lack of temporal granularity means that specific annual events or anomalies affecting efficiency are not captured, limiting the model's utility for detailed, year-specific analyses.

Overall, Model 2 complements the historical analysis by providing an aggregated efficiency estimate, facilitating a broader understanding of Tamweel Bank's long-term operational performance.

Model 3: Comparative Efficiency Analysis of Tamweel Bank and Alizz Islamic Bank (2023)

Model 3 conducts a comparative Data Envelopment Analysis (DEA) between **Tamweel Bank** and **Alizz Islamic Bank** for the year **2023**, providing a direct benchmark to evaluate Tamweel Bank's performance against a prominent peer in the Islamic banking sector.

CCR Efficiency Scores

- **Tamweel Bank (2023): 1.00**
- **Interpretation:** Achieving a perfect CCR efficiency score of **1.00** indicates that Tamweel Bank is operating at the efficiency frontier under Constant Returns to Scale (CCR). This means Tamweel Bank is utilizing its resources optimally relative to its peers, maximizing output without wasting inputs.
- **Alizz Islamic Bank (2023): 0.917**
- **Interpretation:** Alizz Islamic Bank's CCR efficiency score of **0.917** signifies high efficiency but falls short of the optimal frontier estab-

lished by Tamweel Bank. This score suggests that while Alizz Islamic Bank manages its resources effectively, there is still room for improvement to reach the efficiency level of Tamweel Bank.

Comparative Insights

- **Operational Efficiency Gap:**
- **Efficiency Differential:** Tamweel Bank's perfect efficiency score contrasts with Alizz Islamic Bank's score of **0.917**, highlighting a **7.3% efficiency gap**. This difference underscores that Tamweel Bank is more effective in converting inputs into desired outputs compared to Alizz Islamic Bank for the year 2023.
- **Resource Management:**
- **Tamweel Bank:** Demonstrates superior resource management, effectively minimizing input usage while maximizing output generation.
- **Alizz Islamic Bank:** Although highly efficient, Alizz Islamic Bank may benefit from optimizing certain inputs or enhancing output strategies to bridge the efficiency gap with Tamweel Bank.

Strategic Implications for Alizz Islamic Bank

To enhance its operational efficiency and reduce the gap with Tamweel Bank, Alizz Islamic Bank can consider the following strategic focus areas:

- **Operational Cost Management:**
- **Cost Optimization:** Streamlining operational costs without compromising service quality can improve overall efficiency.
- **Process Improvement:** Implementing lean management practices to eliminate waste and enhance productivity.
- **Investment Performance:**
- **Portfolio Diversification:** Optimizing investment portfolios to achieve higher returns while managing risks effectively.
- **Strategic Investments:** Focusing on high-yield investment opportunities that align with the bank's long-term strategic goals.
- **Resource Optimization:**
- **Asset Utilization:** Enhancing the utilization of existing assets to maximize output without incurring additional costs.

- **Liability Management:** Efficiently managing liabilities to ensure financial stability and operational flexibility.

Key Observations:

- **Benchmarking Excellence:**
- **Tamweel Bank as a Benchmark:** Tamweel Bank's perfect efficiency score serves as an exemplary benchmark for Alizz Islamic Bank. By analyzing the practices and strategies that led to Tamweel Bank's optimal performance, Alizz Islamic Bank can identify best practices to emulate.
- **Areas for Improvement:**
- **Targeted Enhancements:** Identifying specific areas where Alizz Islamic Bank lags, such as in certain input efficiencies or output maximization strategies, can guide targeted improvements.
- **Performance Monitoring:** Regularly monitoring efficiency metrics and implementing continuous improvement initiatives can help Alizz Islamic Bank progressively enhance its efficiency scores.

Conclusion of Model 3

The comparative analysis in Model 3 reveals that while both Tamweel Bank and Alizz Islamic Bank exhibit high levels of operational efficiency, Tamweel Bank outperforms its peer by achieving a perfect CCR efficiency score in 2023. This highlights Tamweel Bank's exemplary resource management and operational strategies. Alizz Islamic Bank, with a CCR efficiency score of **0.917**, demonstrates strong performance but also identifies clear opportunities for improvement. By adopting best practices from Tamweel Bank and focusing on strategic areas such as cost management, investment performance, and resource optimization, Alizz Islamic Bank can enhance its operational efficiency and narrow the efficiency gap.

CONCLUSION

This study provides a comprehensive evaluation of operational efficiency within the Islamic banking sector by integrating Data Envelopment Analysis (DEA) with Monte Carlo (MC) simulation. The methodological framework, encompassing both deterministic and stochastic approaches, offers a robust mechanism for assessing and benchmarking the performance of Islamic financial institutions. Through three distinct analyses—historical efficiency assessment, simulated efficiency estimation, and comparative benchmarking—the research elucidates the multifaceted nature of efficiency dynamics in Islamic banking.

Key Findings

1. Historical Efficiency Trends (Model 1):

Dynamic Efficiency Scores: The application of DEA to historical financial data revealed significant variability in efficiency scores over the analyzed period. Peaks in efficiency corresponded with periods of optimal resource utilization and strategic management, while declines indicated emerging operational challenges or adverse external conditions.

Temporal Fluctuations: The study identified specific years where efficiency was maximized, highlighting the impact of effective management practices and favorable economic environments on operational performance. Conversely, periods of reduced efficiency underscored the vulnerabilities and areas requiring strategic interventions.

2. Simulated Efficiency Analysis (Model 2):

Aggregated Efficiency Estimate: Utilizing Monte Carlo simulations, the study generated a synthetic dataset that provided a stable, long-term efficiency estimate. This aggregated score offers a valuable benchmark for understanding overall performance trends, mitigating the influence of year-specific anomalies inherent in historical data.

Error Margin and Insights: The observed 10% discrepancy between simulated and actual efficiency scores underscores the balance between reliability and the inherent uncertainties introduced by stochastic modeling. While simulations enhance the robustness of efficiency estimates, they also highlight the necessity of complementing them with detailed historical analyses.

3. Comparative Efficiency Benchmarking (Model 3):

Benchmarking Against Peers: The comparative analysis between institutions within the Islamic banking sector revealed significant efficiency differentials. By benchmarking against peer institutions, the study identified performance gaps and illuminated best practices that can be emulated to enhance operational efficiency.

Strategic Improvement Areas: The findings emphasized critical areas for strategic improvement, such as operational cost management, investment performance optimization, and resource utilization enhancements. These insights provide actionable guidance for banks aiming to elevate their efficiency standards.

4. Exceptional Performance Indicators:

Identification of Peak Efficiency Factors: The analysis highlighted specific financial and operational factors contributing to peak efficiency years. Factors such as significant profit surges, effective risk management, and strategic adoption of accounting standards played pivotal roles in achieving optimal efficiency scores. Understanding these drivers offers valuable lessons for sustaining high performance.

IMPLICATIONS AND CONTRIBUTIONS

For Islamic Banks

Operational Excellence: The study underscores the importance of effective resource management and strategic operational practices in achieving and maintaining high efficiency levels. Islamic banks can leverage these insights to refine their operational frameworks and enhance competitive positioning.

Strategic Planning and Benchmarking: By adopting a dual-model approach that combines historical and simulated data, banks can engage in more informed strategic planning and benchmarking. This holistic perspective facilitates the identification of both immediate operational challenges and long-term performance trends.

Theoretical Contributions

Methodological Innovation: The integration of DEA with Monte Carlo simulations represents a significant methodological advancement, providing a more nuanced and comprehensive approach to efficiency analysis. This framework can be extended to other sectors and industries, enhancing the versatility and applicability of efficiency evaluations.

Efficiency Evaluation in Islamic Banking: The study contributes to the limited but growing body of literature on efficiency in Islamic banking, offering empirical evidence and methodological insights that can inform future research and practice.

Practical Contributions

Actionable Insights: The findings offer practical recommendations for bank managers and policymakers to optimize operational efficiency. By focusing on identified improvement areas, banks can implement targeted strategies to enhance performance.

Continuous Improvement Initiatives: The study highlights the importance of continuous monitoring and improvement of efficiency metrics. Regular assessments using the proposed framework can help banks maintain optimal performance and adapt to evolving market conditions.

LIMITATIONS AND FUTURE RESEARCH

Data Constraints: The reliance on publicly available financial reports may limit the depth of analysis, as internal factors and qualitative aspects influencing efficiency were not fully captured. Future studies could incorporate more granular data to provide deeper insights.

Simulation Assumptions: While Monte Carlo simulations introduce valuable stochastic elements, the inherent assumptions and simplifications may not fully represent complex operational dynamics. Further research could explore advanced simulation techniques or hybrid models to enhance accuracy.

Comparative Scope: The comparative analysis was limited to specific institutions within a single year. Expanding the scope to include multiple peers and longitudinal comparisons would offer a more comprehensive benchmarking landscape.

Impact of Regulatory Changes: The study touched upon the adoption of new accounting standards and their potential impact on efficiency. Future research could delve deeper into how regulatory and policy changes influence operational performance over time.

Final Reflections

The integration of DEA and Monte Carlo simulations in this study offers a robust and versatile framework for assessing operational efficiency in Islamic banking. By combining historical performance data with simulated scenarios, the research provides a balanced and comprehensive evaluation of efficiency dynamics, capturing both specific temporal trends and overarching performance patterns. The comparative benchmarking further enriches the analysis, highlighting the relative performance of institutions within the sector and identifying strategic opportunities for improvement.

As the banking industry continues to evolve, particularly within the Islamic finance domain, such comprehensive efficiency evaluations will be instrumental in guiding institutions toward sustained operational excellence and competitive advantage. The methodological advancements and empirical findings of this study lay the groundwork for future research and practical applications, fostering a deeper understanding of efficiency mechanisms and their implications for financial institutions.

By addressing both historical and simulated dimensions of efficiency, this study equips Islamic banks with the analytical tools and insights necessary to navigate the complexities of financial management and operational optimization. Embracing a holistic evaluation approach will enable banks to not only enhance their current performance but also build resilient and adaptive operational frameworks for the future.

REFERENCES

Ahmad, I., Nasir, M., & Mokhtar, N. (2021). *Journal of Islamic Accounting and Business Research, 12*(2), 310–327.

Ali, S. M., & Abbas, G. (2022). *International Journal of Islamic and Middle Eastern Finance and Management, 15*(3), 453–470.

Charnes, A., Cooper, W. W., & Rhodes, E. (1978). Measuring the efficiency of decision making units. *European Journal of Operational Research, 2*(6), 429–444. DOI: 10.1016/0377-2217(78)90138-8

Chen, Y., Zhang, X., & Wang, Q. (2021). The Old Boys Club in New Zealand Listed Companies. *Journal of Risk and Financial Management, 14*(8), 345. DOI: 10.3390/jrfm14080342

El-Gamal, M. (2020). *Islamic Finance: Law, Economics, and Practice.* Cambridge University Press.

Gao, Y., Li, S., & Zhou, H. (2023). *Journal of Financial Services Research, 61*(1), 89–112.

Hussain, M., & Imran, M. (2020). *Research in International Business and Finance, 54*, 101263.

Iqbal, M., & Molyneux, P. (2021). *Islamic Finance: Principles and Practice.* Routledge.

Khan, T., & Rahman, S. (2023). *International Journal of Islamic and Middle Eastern Finance and Management, 16*(1), 112–130.

Liu, Y., & Huang, R. (2021). *International Journal of Productivity and Performance Management, 70*(1), 124–142.

Mirakhor, A., & Iqbal, M. (2021). *Islamic Finance: Principles and Practice.* Edward Elgar Publishing.

Mohammad, S., Yasin, M. M., & Ahmad, I. (2023). *Asia Pacific Journal of Management, 40*(1), 211–229.

Nguyen, H. T., & Tran, P. Q. (2021). *Journal of Banking & Finance, 124*, 106038.

Patel, R., & Kumar, S. (2022). *Journal of Financial Services Research, 60*(2), 197–215.

Rahman, S., & Ali, M. (2021). *Strategic Change, 30*(3), 261–275.

Singh, R., & Kaur, H. (2022). *Journal of Islamic Business and Management, 12*(1), 89–105.

Spendolini, M. J. (2021). *The Benchmarking Book.* American Management Association.

Wang, J., Liu, Z., & Yang, X. (2021). *Journal of Quantitative Analysis in Finance, 19*(2), 123–140.

Zhang, L., Wang, Y., & Liu, Q. (2022). *Emerging Markets Finance & Trade, 58*(5), 1342–1360.

Zhou, Y., & Wu, H. (2022). Journal of Islamic Economics. *Banking and Finance, 18*(1), 67–85.

Chapter 5
Performance of Participation Banks in Türkiye

Fikret Kartal
https://orcid.org/0000-0002-2354-8621
Ostim Technical University, Turkey

Nizamülmülk Güneş
Marmara University, Turkey

ABSTRACT

Interest-free banking has been steadily gaining a larger share worldwide, introducing alternative financial products to individuals and businesses. However, since all financial intermediaries operate within the same economic system and the core function of intermediation inherently involves earning profits from borrowing and lending activities, it is not possible to consider interest-free banking as entirely separate from conventional banking. In practice, the differences in terminology and theoretical foundations among banking groups often converge significantly. This study presents the performance of participation banks—the practitioners of interest-free banking in Türkiye—over the past 10 years (December 2014–December 2024). Evaluated alongside the development of deposit banks, the performance results of participation banks highlight their expanding role in the sector. While participation banks have outperformed conventional (deposit) banks in terms of balance sheet growth, their financial results and profitability trends

DOI: 10.4018/979-8-3693-8079-6.ch005

have varied across different years.

INTRODUCTION

In the financial system, interest-free banking represents a significant alternative financial intermediation function, particularly in certain regions and countries. Both interest-based and interest-free finance fundamentally serve the same intermediary role. Financial institutions facilitate the allocation of savings for individuals and companies with surplus funds and meet the financing needs of those requiring funds while also helping them manage various risks.

The core principle of interest-free banking institutions, as reflected in their designation, is the rejection of interest and the implementation of an interest-free banking model. Instead of interest, these banks structure their assets and liabilities based on profit-sharing. Profit is generated through credit transactions linked to trade, partnerships, and investments, while returns on collected funds (deposits) are distributed as profit shares from income pools.

A key point of debate regarding modern interest-free banks is their preference for pre-determined fixed installment credit over true profit-and-loss sharing models. This approach resembles conventional (deposit) banking's interest calculation methods, leading to discussions about its alignment with the fundamental principles of interest-free banking. Specifically, the addition of a profit margin to the transaction cost of trade-based financing—resulting in a fixed, predetermined profit share—raises questions about its consistency with the profit-and-loss-sharing concept. While this study does not focus on the operational principles of interest-free banking in detail, it presents various perspectives related to participation banks.

The terms 'Islamic banking,' 'Islamic Finance,' and 'Interest-Free Banking' are widely used in literature and banking. However, this study primarily uses the terms 'interest-free banking' with reference to global practices, and 'participation banking' specifically within the context of Türkiye. Therefore, in this study, participation banking is referred to as interest-free banking, and conventional banking is referred to as deposit banking. these two banking groups, serving the same customer base within the same economic system, share significant similarities. However, in the context of financial statements, operating results, and performance analysis, participation banking fulfills its intermediation function through 'profit sharing', while conventional banking does so through 'interest'.

This study does not focus on the underlying principles and differences between interest-free banking and conventional banking. The study will briefly address the development of interest-free banks and academic studies on these banks, and then present an analysis of the performance of participation banks in Türkiye compared to conventional banks in the 2015-2024 (10-year) period.

INTEREST-FREE BANKING IN THE WORLD

Interest-free and conventional banking operate in the same economic environment, serving the same customer base and enterprises. Essentially, both banking groups collect funds and finance those in need. Therefore, all these banks, which fulfill the intermediary function, transfer funds from those with surplus funds to those in need of funds.

The significant stages in the development of modern interest-free banking can be summarized as follows (Hussain, Shahmoradi & Turk, 2015; 4-5): Interest-free banking had pioneered practices in Egypt in the early 1960s. The establishment and commencement of operations of the Islamic Development Bank (IDB) in 1975 marked a turning point in terms of interest-free banking, coming just after the establishment of the first modern major commercial Islamic bank in the World, the Dubai Islamic Bank in the United Arab Emirates. The success of this initiative has increased the number of similar banks. Despite the differences between countries, the growth and increasing interest in the sector have brought international regulations for the sector to the agenda. Since 1991, the Accounting and Auditing Organization for Islamic Financial Institutions (AAOIFI) has been issuing accounting, auditing and reporting standards at Islamic financial institutions. The Islamic Financial Services Board (IFSB) was established in 2002 and has been responsible for issuing supervisory and regulatory standards.

The differences between interest-free banks and conventional banks in terms of the products they offer to their customers and, accordingly, the items on their balance sheets can be summarized as follows (Cevik & Charap, 2011; Hanif, 2011; Hasan & Dridi, 2010): Interest-free banking (participation banks in the case of Türkiye) is asset-based and centers on risk sharing while conventional banking is largely debt-based and risk transfer is allowed. When we look at the liability side, iInterest-free and conventional banks collect deposits but the difference emerges in the determination of risk and reward system. Conventional fixed deposit accounts generate predetermined interest earnings on the money while interest-free funds (deposits-participation funds) are accepted through Musharaka and Mudaraba where reward is based on share profit which is variable linked with outcomes of investments made by interest-free banks. At the asset side, both interest-free and conventional banks provide financing to productive channels for reward. The main difference is that conventional banks offer loan for a fixed reward while interest-free banks can charge profit on investments but not charge interest on loans. In a profit loss sharing model, the interest-free banks provide funding to an investment project and receives a pre-determined share of profits. On the other hand, if the project fails or cannot provide the expected return, the bank may lose its investment (loan) entirely or partially.

There is a point to be noted here. Although the profit and loss sharing model is emphasized in theory and discourse for interest-free banks, in practice these banks may turn to financing similar to conventional banks. Although equity participation with risk sharing is at the central, interest-free banks also use debt-like instruments (which is main instrument for commercial banks) that are based on deferred obligation contracts with mark-up financing for payment smoothing similar to conventional credit facilities. The report by the African Development Bank (2011, 15-17) states that the advocates of interest-free finance often express disappointment that interest-free banks provide very little profit sharing (mudaraba) finance and focuses on cost plus trade financing (murabaha) which has many similarities to the loans offered by conventional banks. It is criticized that participation banks also prioritize the demand side and focus less on the supply side.

According to the end-2023 data According to the 2023 year-end data in the Islamic Financial Services Board's (IFSB) report, the interest-free banking sector has an asset size of 2.4 trillion dollars worldwide (IFSB, 2024; 20). The distribution of the aforementioned asset size by region is as follows.

Table 1. Breakdown of interest-free banking in the world by region (2023)

Region	Interest-Free Banking Assets (USD Billion)	Share
Gulf Cooperation Council (GCC)	1.464	62%
Middle East and North Africa [MENA (exc.GCC)	418	18%
East Asia and the Pacific (EAP)	314	13%
South Asia (SA)	84	4%
Europe and Central Asia (ECA)	80	3%
Sub-Saharan Africa (SSA)	13	1%
Total	2.372	

Source: IFSB, 2024; 20.

The breakdown of interest-free banking assets across regions in 2023 shows a globalised and diverse industry. The Gulf region stands out with a share exceeding 60% which reflects its pivotal role in the industry. In the IFSB report, it was stated that the MENA region remains relatively weak in the field of interest-free banking and holds potential for growth. It was emphasized that the Sub-Saharan Africa, Europe and Central Asia, and South Asia regions, which continue to grow, have strong potential despite remaining relatively small (IFSB, 2024; 21). Compound annual growth rates (CAGR) for the period of December 2018-December 2023) shows robust performance and resilience in the face of global challenges. Asset size grew at a CAGR of 10,7% over the stated five-year period while it was 9,9% in financing and 8,8% in deposits (IFSB, 2024; 25). The growth rates for the period 2018-2023 were driven by sectors (the household, real estate, wholesale and construction sectors) that were positively impacted by post-pandemic conditions (IFSB, 2024; 25).

Risks that apply to conventional banks have increased recently, and these risks are naturally relevant to the interest-free banking. The IFSB has addressed several downside risks such as economic volatility and global economic uncertainties that could impact banking operations and profitability (IFSB, 2024; 24). The realization of potential economic risks and downturns could lead to deterioration in asset quality and investment returns, potentially increasing the risk of defaults.

REVIEW OF LITERATURE ON THE PERFORMANCE OF PARTICIPATION BANKS IN TURKİYE

There are numerous academic studies prepared on the activities and performance of participation banks, which fulfill the function of interest-free banking in Türkiye. These studies generally include comparisons with conventional banks.

The research (Coskun, Turanlı, & Yılmaz, 2024; 251), which analyzed funds collected and disbursed by participation banks and covered the period of 2017-2023, stated that participation banks made statistically significant and positive contributions to the country's economy and the banking sector. It was emphasized that participation banks have an important function in bringing the savings of those who are concerned about interest into the economy.

Based on the results of the study covering the period 2010-2021 (Yumurtacı, 2023), it was observed that the CAMELS results of conventional banks were better than those of participation banks, and in recent periods, the CAMELS results of both conventional and participation banks have improved.

A study analyzing the performance of participation banks between 2015-2021 stated that the financial performance of participation banks showed both positive and negative changes. The progress of participation banks in the banking sector was observed to be slow (Yıldız, 2023).

The study (Gürçay & Dağıdır, 2022), which investigated the response of participation and deposit banks to the COVID-19 pandemic, concluded that participation banks were more efficient in terms of asset quality during the pandemic, and that there was no significant difference in other variables (profitability, risk and solvency, liquidity, and capital adequacy).

In a study (Elmas & Yetim, 2021) that comparatively examined the financial performance of participation banks on an international scale for the period of 2012-2019, it was found that the countries' general performance rankings and the global Islamic banking total asset share rankings were exactly the same. Türkiye ranked 5th in both rankings.

In an analysis comparing private and public participation banks, it was stated that private participation banks were clearly ahead of public participation banks in terms of the number of personnel, the number of branches, net profit, total assets and the efficiency of the funds provided (Eke & Sevinç, 2021).

According to the data obtained in the study analyzing the 2015-2019 period (Yılmaz & Özgür, 2021), it was determined that a one-unit increase in the capital adequacy ratio of participation banks reduced asset profitability by 13%, while a one-unit increase in the funds loaned/total assets ratio reduced asset profitability by 291%. The negative impact of the increase in the capital adequacy ratio on return on assets is attributed to the minimization of risk, while the negative impact of the increase in the ratio of funds used/ total assets is attributed to the increase in the rate of funds used leading to an increase in non-performing loans.

In the results of the study (Baykuş & Bektaş, 2021), which aimed to identify the factors affecting the profit-sharing ratios of participation banks in the 2012-2020 period, a two-way causality was found between profit-sharing ratios and conventional banking one-month deposit interest, and a one-way causality relationship was found from the weighted average funding cost to profit-sharing ratios. No causal relationship was found between the profit-sharing rate, exchange rate, and consumer price index.

In the study using data from the 2011-2016 period (Hamarat, 2024; 73-74), it was stated that deposit and participation banks operated with high foreign resources and low equity, and that this situation harmed bank efficiency in the event of an increase in borrowing costs. It was also stated that high inflation and interest rates negatively affected efficiency. It was noted that conventional banks had higher cost efficiency, except for the years 2011 and 2012.

In another study (Batir, Volkman, & Gungor, 2017) examining the period 2005-2013 stated that the efficiency in participation banks was positively affected by expense, loans and non-performing loans while negatively affected by profit, bank size, GDP growth and inflation. The figures on participation and deposit banks (seperately state, foreign, and domestic private banks) indicate that participation banks have higher average technical, allocative and cost efficiency.

In a study (Karapinar & Dogan, 2015; 31) that analyzed the ratios of conventional and participation banks operating in Türkiye between 2006 and 2011, it was stated that the two bank groups achieved different results in terms of management, liquidity and sensitivity. Compared to the conventional banks, the participation banks performed well from the viewpoint of their sensitivity to the market risks while they have done poorer with respect to liquidity and management. When the pros and cons are evaluated together, the study in question concluded that participation banks are as effective as conventional banks.

The study (Aysan, Dolgun, & Turhan, 2013; 110), which analyzed the situation and potential of participation banks, highlighted the room for growth potential for participation banks in Turkiye. The reasons given for this are the regulations made for the sector, new players in the sector, and new products.

THE DEVELOPMENT OF PARTICIPATION BANKS IN TÜRKİYE

The first regulation regarding interest-free banking in Türkiye was made with a decree in 1983, and the first institutions were established in 1985 through 'special finance houses'. The aim of Turkiye was to attract more foreign direct investment from the Arab Gulf states (IMF, 2017; 83). In 1999, special finance houses were brought under the scope of the banking law while preserving their principles. In 2005, with the new banking law numbered 5411, the name of these institutions was changed to 'participation banks'. Participation banks have the same conditions as conventional deposit banks in the areas of regulation and supervision, and within the same legal regulations, there are some special provisions specific to them. The deposit accounts (participation funds) opened in these banks are placed under state guarantee within a certain limit, as in conventional banks.

The regulations made in the banking law in Türkiye, changing the name of private financial institutions to participation banks, introducing deposit guarantee for these institutions, and subjecting them to the same regulations as banks have caused controversy in some respects. These regulations aim to increase trust in participation banks and thereby raise their sector share, while on the other hand, it is stated that the distinction between participation and conventional banks has weakened. The fact that the asset structure of participation banks predominantly shares the same characteristics as installment-based corporate and individual loans in commercial banks has essentially highlighted the similarities between these banks and commercial banks. Profit-sharing based financing, which is constantly emphasized as a feature of these institutions, is practically non-existent in the balance sheets of participation banks. Therefore, it can be stated that the conceptual differences expressed regarding participation banks are not reflected in practice.

This situation has also been the subject of several studies. There are some studies founding that interest-free banks have similiar structures to conventional banks. The point that is emphasized is that participation banks generally turn to debt-based financing (nearly all of their loans) used by conventional banks and do not prefer profit/loss sharing-based financing, which is the foundation of interest-free banking. For this reason, it is stated that participation banks practically finance trade, not commercial enterprises, and as a result, the interest-free loan calculations of participation banks and the interest-based loan calculations of conventional banks show considerable similarity (Geçer, 2023). The main criticism is that interest-free banks mainly provide short term trade finance which is the same as conventional banks offer and their operations do not conform to the principle of profit and loss sharing. On the other hand, the returns on the interest-free deposit accounts are effectively pegged to the returns on conventional banking deposits because of competitive reasons and economic structure in which banks and all units operate (Kartal, 2012; 199). African Development Bank (2011, 15-17) states that interest-free banks provide very little profit sharing (mudaraba) finance and focuses on cost plus trade financing (murabaha) which has many similarities to the loans offered by conventional banks.

Below is a demonstration of the change in the portion of sector assets held by participation banks, which were renamed in 2005 and have since grown rapidly.

Table 2. Development of participation banks

Banking sector assets	2005	2010	2020	2024	CAGR
Asset Size (Million USD)	292.473	636.499	773.847	876.134	5,9%
Participation Banks	7.410	28.073	59.264	75.693	13,0%
Conventional Deposit Banks	285.063	608.426	714.583	800.441	5,6%
Participation Banks' Share (in Deposit + Participation Banks)	2,5%	4,4%	7,7%	8,6%	

Source: BRSA, 2025, Monthly Banking Sector Data (Basic Analysis), https://www.bddk.org.tr/BultenAylik/en

The participation banks' sector share, which was 2.5%, increased to the level of 8.6% by the end of 2024. In the 19-year period between December 2005 and December 2024, participation banks achieved an average annual growth (CAGR) of 13% in dollar terms, while conventional banks had a growth of 5.6%.

Over the past decade, newly founded state-owned participation banks have played a crucial role in the sector's expansion. Ziraat Participation Bank was established in 2015, followed by Vakıf Participation Bank in 2016 and Türkiye Emlak Participation Bank in 2019. These new public participation banks, launched in line with the government's supportive policies, have helped reach a broader customer base and contributed to increasing the share of participation banking within the overall banking sector.

As of January 2025, a total of nine participation banks—including three state-owned and two digital banks—are operating with 1,513 branches. These banks employ 21,358 personnel. Meanwhile, Türkiye's banking sector also includes 35 deposit banks, three of which are state-owned, employing a total of 180,322 staff.

The positive approach of public authorities towards participation banking has created a favorable environment for these banks. Despite its growing popularity, participation banks faces strong competition from the conventional banks (IMF, 2017; 83). The advantages that conventional banks have in areas such as scale, product diversity, and digital banking challenge participation banks in competition.

ANALYSIS OF THE PERFORMANCE OF TURKISH PARTICIPATION AND DEPOSIT BANKS IN THE 2015-2024 PERIOD

The aim of this study is to analyze and compare the performance of conventional deposit banks and participation banks over a 10-year period from the end of 2015 to the end of 2024. For this purpose, data published by the Banking Regulation and Supervision Authority (BRSA) have been used. The study examines the consolidated balance sheets and income statements of both bank groups, as well as the ratios calculated based on these financial figures. The analysis section consists of two subsections: key balance sheet items and fundamental profitability indicators. All values have been considered in terms of U.S. dollars.

The equivalents of some key items in the financial statements of deposit (conventional) banks and participation banks are shown below.

Table 3. Equivalents of key financial items in participation banks

Conventional Deposit Banks	Interest-Free Participation Banks
Deposit	Participation Fund
Demand Deposit	Special Current Account
Term Deposit	Participation Accounts
Interest Income	Profit Share Income
Interest Income from Loans	Profit Share Income from Loans
Interest Expenses	Profit Share Expenses

One of the critical dates in evaluating the last 10 years (2014–2024) of the Turkish banking sector is the year 2020, when the pandemic crisis began. In response to economic developments in 2020, expansionary monetary and fiscal policies were implemented. Subsequently, from September 2021 onward, an unconventional monetary policy was adopted, severing the link between inflation and interest rates, and a highly negative real interest rate was used as a tool.

In light of the destruction caused by an environment of high inflation and reduced predictability, a gradual transition to conventional monetary policy began in June 2023. High positive real interest rates and a strong Turkish lira have become the key instruments of the new period.

Development of Balance Sheet Items

For the analysis of the development of balance sheet items in banks during the 2015-2024 period, it has been considered beneficial to make a dual distinction: (1) the period from 31.12.2014 to 31.12.2024 (10 years) and (2) the period from 31.12.2019 to 31.12.2024 (5 years), starting before the pandemic crisis. Below, the development of key balance sheet items and period profits by year, along with the compound annual growth rate (CAGR), are presented. The data are provided in USD terms.

The fluctuating nature of the Turkish Lira, particularly the high inflation and appreciation observed after 2021, has led to fluctuations and significant changes in the USD equivalent of balance sheet items.

The total assets of deposit banks, which stood at USD 778 billion in 2014, declined to USD 659 billion in 2019 and reached USD 800 billion by the end of 2024. Over the 10-year period between the end of 2014 and the end of 2024, the CAGR for total assets was 0.3%, while cash and cash equivalents grew at a significantly high CAGR of 8.3%. Securities recorded a positive growth rate of 0.7%, whereas loans saw a negative CAGR of -2.1%. On the liabilities side, the average annual growth rate for deposits was 1.4%, while other key items experienced a decline.

However, when analyzing the last five-year period, deposit banks achieved the highest growth in liquid assets, with a CAGR of 4% (an annual average growth of 10%), while loans remained at the same level and securities recorded a relatively high growth rate of 5.7%. Deposits had a CAGR of 4.2%, borrowings from other banks grew at 2.3%, and issued securities exhibited a negative CAGR of -2.3%.

Table 4. Development of main balance sheet items of deposit and participation banks

Deposit Banks	Sizes as of Year-End (31.12.xx)							CAGR	
Million USD	2014	2019	2020	2021	2022	2023	2024	2015-2024	2019-2024
Cash and Cash Equivalents	48.487	67.089	64.799	80.574	76.402	97.572	107.810	8,3%	10,0%
Loans (Performing)	483.037	389.928	418.271	314.110	348.801	339.366	390.985	-2,1%	0,1%
Securities	124.958	101.267	123.796	98.966	111.573	118.477	133.387	0,7%	5,7%
Total Assets	778.485	659.178	714.583	596.456	660.389	687.031	800.441	0,3%	4,0%
Deposit	426.259	397.369	424.253	359.429	426.726	454.592	487.853	1,4%	4,2%
a) Demand Deposits	78.778	95.037	132.918	135.945	148.274	148.538	161.953	7,5%	11,2%
b) Term Deposits	347.481	302.332	291.335	223.484	278.452	306.054	325.900	-0,6%	1,5%
Securities Issued (Net)	34.780	27.420	25.071	18.837	13.036	14.427	22.790	-4,1%	-3,6%
Payables to Banks	110.545	72.514	71.185	62.685	61.158	65.143	81.252	-3,0%	2,3%
Equity	86.547	71.839	70.198	46.477	65.614	63.201	68.549	-2,3%	-0,9%
Profit of the Period	9.890	6.903	6.590	5.895	20.353	17.951	14.423	3,8%	15,9%
Participation Banks	**Sizes as of Year-End (31.12.xx)**							**CAGR**	
Million USD	**2014**	**2019**	**2020**	**2021**	**2022**	**2023**	**2024**	**2015-2024**	**2019-2024**
Cash and Cash Equivalents	5.346	8.838	6.832	7.896	7.791	10.045	12.944	9,2%	7,9%
Loans (Performing)	27.580	22.930	30.151	25.500	30.968	30.784	33.600	2,0%	7,9%
Securities	2.906	7.002	10.511	9.382	11.248	12.924	10.538	13,7%	8,5%
Total Assets	44.911	47.884	59.264	54.417	63.548	69.318	75.693	5,4%	9,6%
Deposit	28.078	36.266	43.568	42.198	47.681	51.533	50.620	6,1%	6,9%
a) Demand Deposits	6.882	12.157	19.695	19.740	19.952	19.786	20.328	11,4%	10,8%
b) Term Deposits	21.196	24.109	23.873	22.458	27.729	31.748	30.292	3,6%	4,7%
Securities Issued (Net)	0	0	0	0	0	0	0	-	-
Payables to Banks	6.422	2.183	3.020	2.078	1.960	2.824	4.304	-3,9%	14,5%
Equity	4.165	3.664	3.745	2.753	4.506	4.886	6.077	3,8%	10,6%
Profit of the Period	61	410	504	414	1.588	1.809	1.794	40,2%	34,3%

Source: BRSA, Monthly Banking Sector Data (Basic Analysis), https://www.bddk.org.tr/BultenAylik/en

Participation banks achieved higher CAGR rates in all major balance sheet items compared to deposit banks over the 2014-2024 period. They recorded a CAGR of 5.4% in total assets, 13.5% in securities, 2% in loans, 6.1% in deposits, and 3.8% in equity, indicating a stronger performance compared to deposit banks. When considering only the last five years, participation banks again demonstrated higher growth rates, with a CAGR of 9.6% in total assets, 7.9% in loans, 8.5% in securities, 6.9% in deposits, and 14.5% in borrowings from other banks. Participation banks did not issue securities.

Due to their stronger performance in balance sheet growth, participation banks also increased their market share within the banking sector. This trend is evident from the year-end (31.12.xx) ratios presented below.

Table 5. The share of participation banks' main balance sheet items in the sector

Participation Banks' Share (in Deposit + Participation Banks)	2014	2019	2020	2021	2022	2023	2024
Cash and Cash Equivalents	9,9%	11,6%	9,5%	8,9%	9,3%	9,3%	10,7%
Loans (Performing)	5,4%	5,6%	6,7%	7,5%	8,2%	8,3%	7,9%
Securities	2,3%	6,5%	7,8%	8,7%	9,2%	9,8%	7,3%
Total Assets	5,5%	6,8%	7,7%	8,4%	8,8%	9,2%	8,6%
Deposit (Participation Funds)	6,2%	8,4%	9,3%	10,5%	10,1%	10,2%	9,4%
a) Demand Deposits	8,0%	11,3%	12,9%	12,7%	11,9%	11,8%	11,2%
b) Term Deposits	5,7%	7,4%	7,6%	9,1%	9,1%	9,4%	8,5%
Securities Issued (Net)	0,0%	0,0%	0,0%	0,0%	0,0%	0,0%	0,0%
Payables to Banks	5,5%	2,9%	4,1%	3,2%	3,1%	4,2%	5,0%
Equity	4,6%	4,9%	5,1%	5,6%	6,4%	7,2%	8,1%

Source: Compiled from the data in Table 1. (BRSA, Monthly Banking Sector Data (Basic Analysis), https://www.bddk.org.tr/BultenAylik/en)

In 2014, participation banks accounted for 5.5% of total banking assets. This share increased to 6.8% in 2019, 9.2% in 2023, and 8.6% in 2024. During the 2014-2024 period, their share of total loans increased from 5.1% to 7.9%, their share in securities portfolios rose from 2.3% to 7.3%, and their share of total deposits grew from 5.9% to 9.4%. Thus, the strong performance of participation banks, as outlined in the previous table, has contributed to their growing position in the banking sector.

The weight of deposit and participation banks' assets and liabilities within their balance sheets can be observed through the following ratios.

Table 6. Balance sheet distribution of deposit and participation banks

Deposit Banks	Year End (31.12.xx) Ratios						
%	2014	2019	2020	2021	2022	2023	2024
Loans / T. Assets	62,0%	59,2%	58,5%	52,7%	52,8%	49,4%	48,8%
Securities / T. Assets	16,1%	15,4%	17,3%	16,6%	16,9%	17,2%	16,7%
Cash and Cash Equiv. / T. Assets	6,2%	10,2%	9,1%	13,5%	11,6%	14,2%	13,5%
Deposit / T. Liabilities	54,8%	60,3%	59,4%	60,3%	64,6%	66,2%	60,9%
Demand Deposit / Total Deposit	18,5%	23,9%	31,3%	37,8%	34,7%	32,7%	33,2%
Equity / Total Liabilities	11,1%	10,9%	9,8%	7,8%	9,9%	9,2%	8,6%
Participation Banks	**Year End (31.12.xx) Ratios**						
%	2014	2019	2020	2021	2022	2023	2024
Loans / T. Assets	61,4%	47,9%	50,9%	46,9%	48,7%	44,4%	44,4%
Securities / T. Assets	6,5%	14,6%	17,7%	17,2%	17,7%	18,6%	13,9%
Cash and Cash Equiv. / T. Assets	11,9%	18,5%	11,5%	14,5%	12,3%	14,5%	17,1%
Deposit (Participation Funds) / T. Liabilities	62,5%	75,7%	73,5%	77,5%	75,0%	74,3%	66,9%
Demand Deposit / Total Deposit (Participation Funds)	24,5%	33,5%	45,2%	46,8%	41,8%	38,4%	40,2%
Equity / Total Liabilities	9,3%	7,7%	6,3%	5,1%	7,1%	7,0%	8,0%

Source: Compiled from the data in Table 1. (BRSA, Monthly Banking Sector Data (Basic Analysis), https://www.bddk.org.tr/BultenAylik/en)

At the end of 2014, deposit and participation banks had similar 'Loans to Assets' ratios of 62%. By the end of 2024, this ratio had decreased to 48.8% for deposit banks and 44.4% for participation banks. The declining share of loans in participation banks' consolidated balance sheets has been replaced by an increasing share of securities and liquid assets. While deposit banks slightly increased the share of securities in their assets, they experienced a significant increase in liquid assets. The share of deposits (participation funds), particularly demand deposits, has increased in both banking groups. However, participation banks have a more pronounced share of deposits on the liabilities side. In both banking groups, the share of equity in liabilities has declined, with deposit banks experiencing a more significant decline.

Key Profitability Indicators

The primary activity of the banking sector is to collect funds and allocate them into various assets, primarily loans. Therefore, the main role of banks is to act as intermediaries between surplus and deficit sectors. The primary objective of this intermediation process is to generate profit. Banks' main sources of income and expenses are interest income and interest expenses (profit share), respectively, with the ultimate goal being to achieve high asset quality, strong equity capital, and high return on equity (RoE).

Key ratios related to interest income/expenses (profit share), net interest margin (net profit margin), and return on equity (RoE) for deposit and participation banks are illustrated in the graphs below. Since income and profit figures are spread throughout the year, the asset, liability, and equity values used in the calculation of these ratios have been averaged based on their values at the end of each month within the respective year.

The high levels of interest (profit share) and profitability ratios, particularly after 2022, are mainly due to the appreciation of the Turkish Lira in a high-inflation environment. For example, in 2024, while the USD exchange rate increased by 18%, the year-end Consumer Price Index (CPI) inflation rate was 44%, and the annual average CPI inflation rate was 58.5%. This led to an appreciation of the Turkish Lira. Although high nominal interest rates (profit share) and high nominal profitability ratios resulted in negative real returns in TL terms due to inflation, they generated extremely high returns in USD terms.

Figure 1. Interest (profit share) income on assets

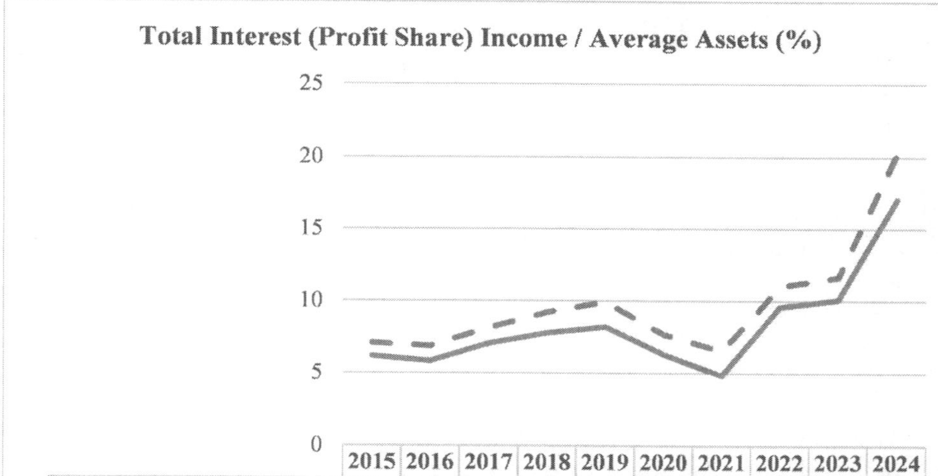

Total Interest (Profit Share) Income / Average Assets (%)

	2015	2016	2017	2018	2019	2020	2021	2022	2023	2024
Deposit Banks	7,1	6,9	8,1	9,2	9,9	7,7	6,5	11,0	11,6	20,3
Participation Banks	6,2	5,8	7,0	7,8	8,2	6,3	4,8	9,6	10,1	17,0

Source: *Compiled from BRSA data (BRSA, Monthly Banking Sector Data (Advanced Analysis), https://www.bddk.org.tr/BultenAylik/en/Home/Gelismis)*

The return on total interest income (profit share) as a percentage of average assets has always been higher for deposit banks. In recent years, the appreciation of the Turkish Lira and high inflation have increased yields. Since both deposit and participation banks operate in the same economic environment, and since interest not only represents the cost of money but also plays a fundamental role in market equilibrium and financial intermediation, the yields of both banking groups have shown a parallel trend.

The yield on loans, calculated as the ratio of interest (profit share) income from loans to average loans, is illustrated in the graph below.

Figure 2. Interest (profit share) income from performing loans

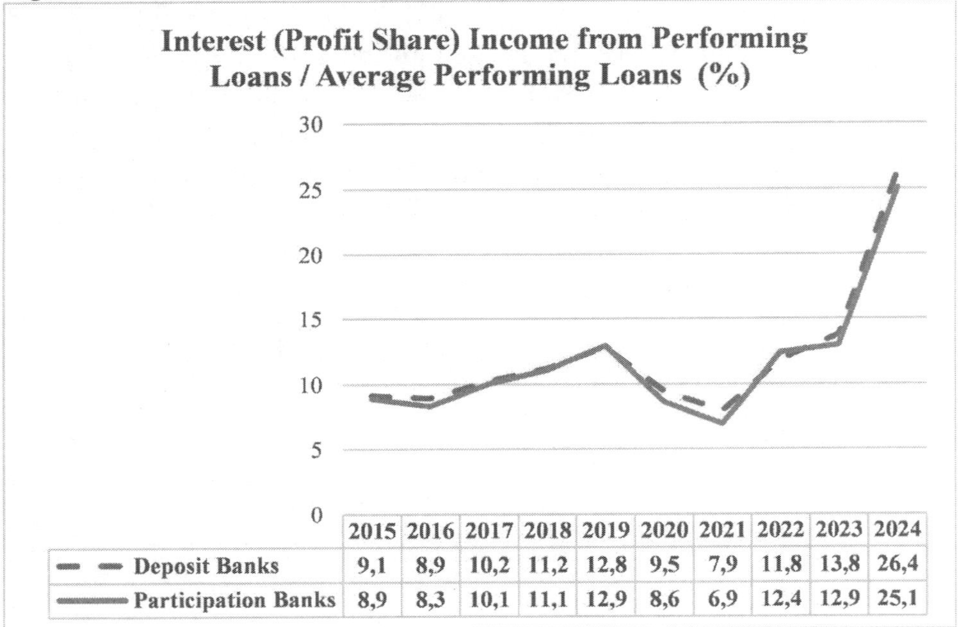

	2015	2016	2017	2018	2019	2020	2021	2022	2023	2024
Deposit Banks	9,1	8,9	10,2	11,2	12,8	9,5	7,9	11,8	13,8	26,4
Participation Banks	8,9	8,3	10,1	11,1	12,9	8,6	6,9	12,4	12,9	25,1

Source: *Compiled from BRSA data (BRSA, Monthly Banking Sector Data (Advanced Analysis), https://www.bddk.org.tr/BultenAylik/en/Home/Gelismis)*

The interest (profit share) income earned from loans by deposit and participation banks has moved in parallel, with deposit banks generally having slightly higher yields. Thus, given that both banking groups operate under the same economic conditions, they naturally demand similar return levels for loans.

The ratio of interest (profit share) expenses paid by banks on borrowed funds to total liabilities for the 2015-2024 period is shown below.

Figure 3. Interest expense (profit share) incurred for liabilities (excluding equity)

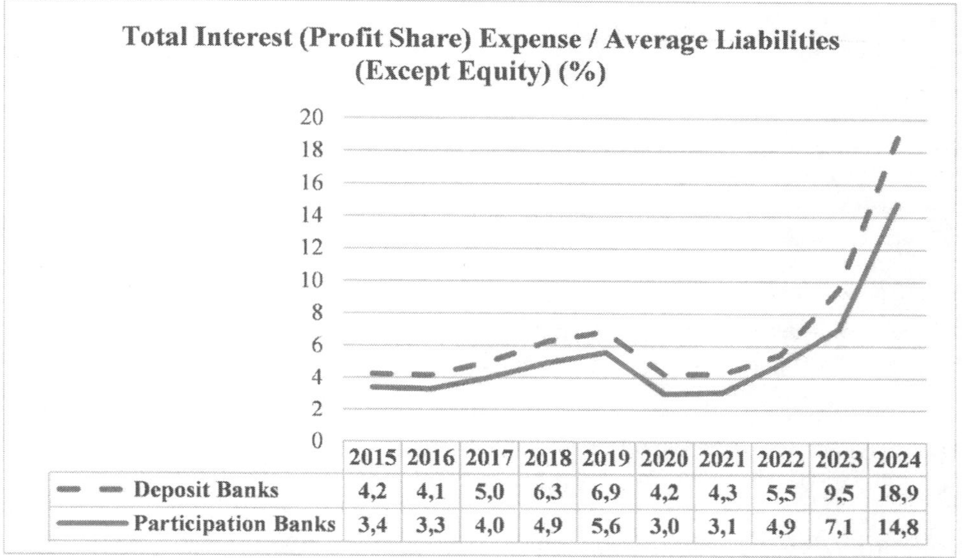

	2015	2016	2017	2018	2019	2020	2021	2022	2023	2024
Deposit Banks	4,2	4,1	5,0	6,3	6,9	4,2	4,3	5,5	9,5	18,9
Participation Banks	3,4	3,3	4,0	4,9	5,6	3,0	3,1	4,9	7,1	14,8

Source: *Compiled from BRSA data (BRSA, Monthly Banking Sector Data (Advanced Analysis), https://www.bddk.org.tr/BultenAylik/en/Home/Gelismis)*

Participation banks pay lower profit shares (interest expenses) on their liabilities compared to deposit banks. Consequently, while participation banks generate lower returns from their assets, they also bear lower funding costs for their liabilities.

The development of the net interest margin (profit share margin), which is calculated by expressing the difference between interest income and interest expenses (profit share) as a percentage of total assets, is shown below.

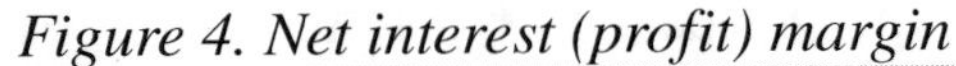

Figure 4. Net interest (profit) margin

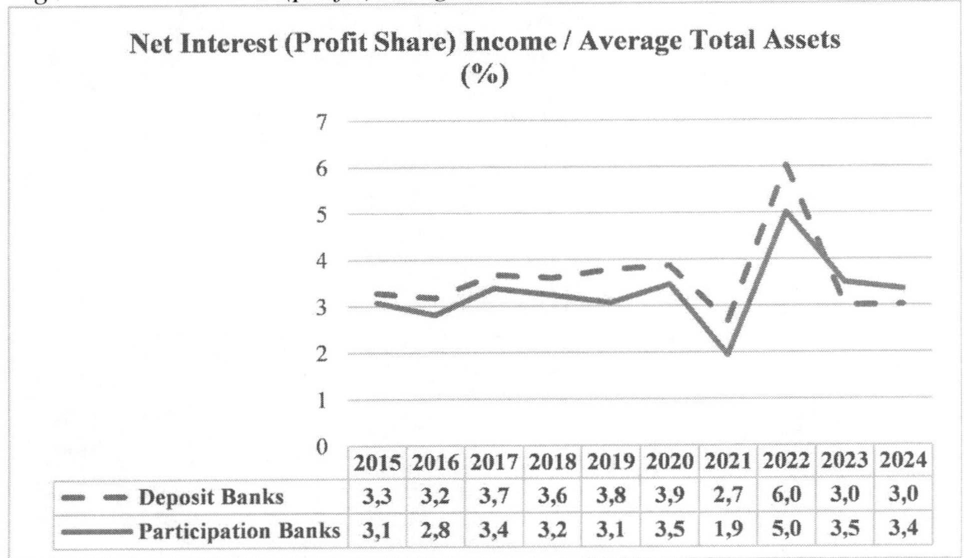

	2015	2016	2017	2018	2019	2020	2021	2022	2023	2024
Deposit Banks	3,3	3,2	3,7	3,6	3,8	3,9	2,7	6,0	3,0	3,0
Participation Banks	3,1	2,8	3,4	3,2	3,1	3,5	1,9	5,0	3,5	3,4

Source: *Compiled from BRSA data (BRSA, Monthly Banking Sector Data (Advanced Analysis), https://www.bddk.org.tr/BultenAylik/en/Home/Gelismis)*

Until 2022, deposit banks had a higher net interest margin (profit share margin). However, in 2023 and 2024, participation banks achieved higher margins. Due to the sharp rise in inflation in 2022, interest income increased more than interest expenses, causing the margin to rise above its usual levels. Under normal conditions, banks' net interest (profit share) margin is in the range of 3-3.5%.

A critical indicator of bank performance and overall financial results, return on equity (RoE), for both banking groups during the 2015-2024 period is as follows:

Figure 5. Return on equity (RoE)

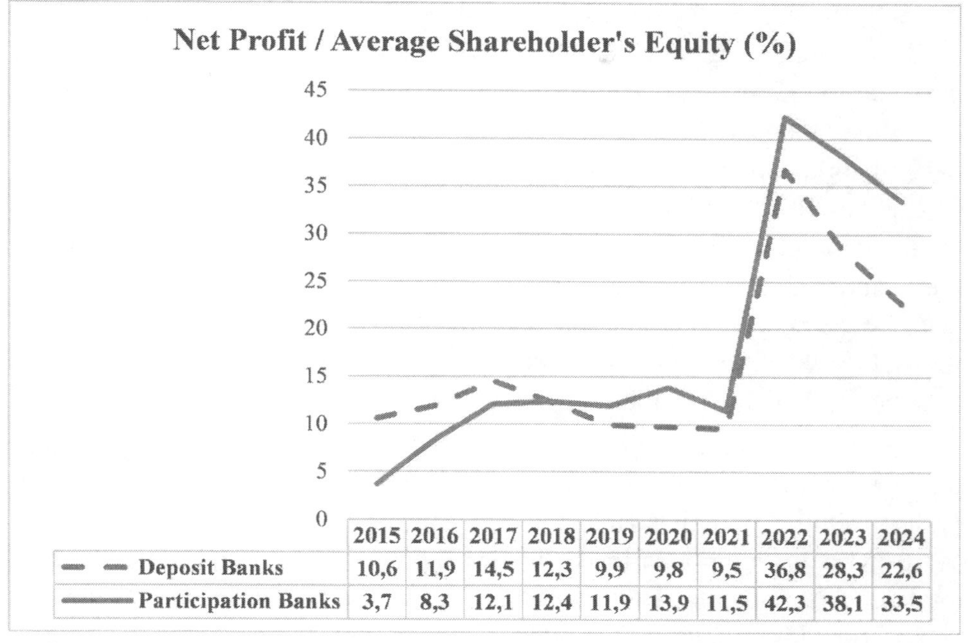

	2015	2016	2017	2018	2019	2020	2021	2022	2023	2024
Deposit Banks	10,6	11,9	14,5	12,3	9,9	9,8	9,5	36,8	28,3	22,6
Participation Banks	3,7	8,3	12,1	12,4	11,9	13,9	11,5	42,3	38,1	33,5

Source: *Compiled from BRSA data (BRSA, Monthly Banking Sector Data (Advanced Analysis), https://www.bddk.org.tr/BultenAylik/en/Home/Gelismis)*

From 2015 to 2017, deposit banks had higher RoE, whereas participation banks have maintained a higher RoE since 2018. The high inflation and the appreciating Turkish Lira observed since 2022 have led to significantly higher RoE ratios in USD terms in recent years.

CONCLUSION

This study aims to present the performance of participation banks over the 2015-2024 period (10 years) alongside the development of deposit banks. It utilizes the consolidated balance sheets and income statements of participation and deposit banks, evaluating data in USD terms. The examined period is divided into two sub-periods: the last 10 years and the last five years, starting just before the pandemic (December 2019–December 2024). Since the data is presented in USD, the high volatility of the USD/TRY exchange rate directly impacts the results. In particular, the recent appreciation of the Turkish Lira has significantly increased USD-denominated returns.

Based on the compound annual growth rate (CAGR), participation banks have exhibited stronger balance sheet performance compared to deposit banks over the last 10 years. Between December 2014 and December 2024, participation banks recorded a CAGR of 5.4% in total assets, 13.5% in securities, 2% in loans, 6.1% in deposits, and 3.8% in equity. In contrast, deposit banks recorded a CAGR of 0.3% in total assets, 0.7% in securities, -2.1% in loans, and 8.3% in cash and cash equivalents. When considering only the last five years, participation banks again demonstrated higher growth rates, with a CAGR of 9.6% in total assets, 7.9% in loans, 8.5% in securities, 6.9% in deposits, and 14.5% in borrowings from other banks.

Due to this stronger balance sheet performance, the market share of participation banks in total assets increased from 5.5% to 8.6% over the 10-year period. Similarly, their share of total loans rose from 5.1% to 7.9%, their share in securities portfolios grew from 2.3% to 7.3%, and their share of total deposits expanded from 5.9% to 9.4%. At the end of 2014, both deposit and participation banks had similar 'Loans to Assets' ratios of 62%. By the end of 2024, this ratio had declined to 48.8% for deposit banks and 44.4% for participation banks.

The return on total interest income (profit share) as a percentage of average assets has always been higher for deposit banks. In recent years, the appreciation of the Turkish Lira and high inflation have boosted yields. While participation banks generate lower returns from their assets, they also incur lower funding costs for their liabilities. The interest (profit share) income earned from loans by deposit and participation banks has moved in parallel, with deposit banks generally achieving slightly higher yields. Since both banking groups operate under the same economic conditions, they naturally demand similar return levels for loans.

Until 2022, deposit banks had a higher net interest margin (profit share margin). However, in 2023 and 2024, participation banks achieved higher margins. From 2015 to 2017, deposit banks maintained a higher return on equity (RoE), whereas participation banks have held a higher RoE since 2018.

Looking at the last 10 years as a whole, the increased presence of state-owned players in participation banking and relatively strong performance have contributed to the rising market share of participation banks. On the other hand, deposit banks maintain competitive advantages due to their larger scale, widespread branch networks, broad customer base, and significant investments in new technologies, making competition more challenging for participation banks.

REFERENCES

African Development Bank. (2011). *Islamic Banking and Finance in North Africa Past Development and Future Potential.*

Aysan, A. F., Dolgun, M. H., & Turhan, M. İ. (2013). Assessment of the Participation Banks and Their Role in Financial Inclusion in Turkey. *Emerging Markets Finance & Trade*, *49*(5), 99–111. DOI: 10.2753/REE1540-496X4905S506

Batir, T. E., Volkman, D. A., & Gungor, B. (2017). Determinants of Bank Efficiency in Turkey: Participation Banks Versus Conventional Banks. *Borsa Istanbul Review*, *17*(2), 86–96. DOI: 10.1016/j.bir.2017.02.003

Baykuş, O., & Bektaş, S. (2021, August). Katılım Bankalarının Kâr Paylaşım Oranlarını Belirleyen Etmenler Üzerine Ampirik Bir İnceleme: Türkiye Katılım Bankaları Örneği. *The Journal of Accounting and Finance*, (Special Issue), 397–422.

BRSA. (2025). Monthly Banking Sector Data (Basic Analysis), https://www.bddk.org.tr/BultenAylik/en

Çevik, S., & Charap, J. (2011). *The Behavior of Conventional and Islamic Bank Deposit Returns in Malaysia and Turkey.* IMF Working Paper, WP/11/156.

Coskun, Ş., Turanlı, M., & Yılmaz, K. (2024). The Relationship of Funds Collected and Disbursed in Participation Banks with Sector Shares and the Effect of Macroeconomic Indicators. *Journal of Islamic Research*, *35*(2), 240–253.

Eke, V., & Sevinç, H. (2021). Türkiye'deki Katılım Bankalarının Etkinlik Analizi: Özel ve Kamu Bankalarının Karşılaştırılması. *Iğdır Üniversitesi Sosyal Bilimler Dergisi*, (28), 434–451.

Elmas, B., & Yetim, A. (2021). Katılım Bankalarının Finansal Performanslarının TOPSIS Yöntemi İle Uluslararası Boyutta Değerlendirilmesi. *Uluslararası İslam Ekonomisi ve Finansı Araştırmaları Dergisi*, *7*(3), 230–263.

Gürçay, H. R., & Dağıdır, C. (2022). COVID-19 Sürecinde Katılım Bankaları ile Özel Mevduat Bankalarının Performans Değerlendirmesi: Türkiye Örneği. *Uluslararası Finansal Ekonomi ve Bankacılık Uygulamaları Dergisi*, *3*(1), 1–25.

Hamarat, Ç. (2024). The Efficiency of Participation and Conventional Banking in Turkiye: A Stochastic Frontier Approach. *Toplum Ekonomi ve Yönetim Dergisi*, *5*(1), 56–79. DOI: 10.58702/teyd.1341253

Hanif, M. (2011). Differences and Similarities in Islamic and Conventional Banking. *International Journal of Business and Social Science*, *2*(2), 166–175.

Hasan, M., & Dridi, J. (2010). *The Effects of the Global Crisis on Islamic and Conventional Banks: A Comparative Study*. IMF Working Paper, WP/10/201.

Hussain, M., Shahmoradi, A., & Turk, R. (2015). *An Overview of Islamic Finance*. IMF Working Paper, WP/15/120.

IFSB. (2024). *Islamic Financial Services Industry Stabiliıty Report 2024*. https://www.ifsb.org/wp-content/uploads/2024/09/IFSB-Stability-Report-2024-8.pdf

IMF. (2017). *Ensuring Financial Stability in Countrıes with Islamic Banking-Country Case Studies*. IMF Country Report No. 17/145.

Karapinar, A., & Dogan, İ. C. (2015). An Analysis on the Performance of the Participation Banks in Turkey. *Accounting and Finance Research*, *4*(2), 24–33. DOI: 10.5430/afr.v4n2p24

Kartal, F. (2012). Interest-Free Banking in the World and a Financial Analysis of the Turkey Experience. *International Research Journal of Finance and Economics*, *93*, 183–201.

Yıldız, N. (2023). Türkiye'de Faaliyet Gösteren Katılım Bankalarının Performans Analizi. *Cumhuriyet Üniversitesi İktisadi ve İdari Bilimler Dergisi*, *24*(1), 36–49. DOI: 10.37880/cumuiibf.1173166

Yılmaz, C., & Özgür, E. (2021). Katılım Bankalarında Kârlılığa Etki Eden Faktörlerin Tespiti İçin Panel Veri Analizi Uygulaması. *Muhasebe Ve Finansman Dergisi*, (92), 1–20. DOI: 10.25095/mufad.944461

Yumurtacı, R. (2023). Türkiye'deki Konvansiyonel Bankalar ile Katılım Bankalarının CAMELS Analizi ile Karşılaştırılması. *Süleyman Demirel Üniversitesi Vizyoner Dergisi*, *14*(39), 1077–1097. DOI: 10.21076/vizyoner.1196650

Chapter 6
The Expansion of Islamic Fintech:
The Digital Transformation of Financial Services in the World and Turkey

Burak Aktürk
https://orcid.org/0000-0001-9210-6985
Marmara University, Turkey

Yunus Emre Gürbüz
https://orcid.org/0000-0003-2355-1852
Istanbul University, Turkey

Yusuf Sait Turkan
https://orcid.org/0000-0001-7240-183X
Istanbul University-Cerrahpasa, Turkey

ABSTRACT

Fintech is a concept that encompasses new technologies and financial innovations with the objective of optimizing the delivery and utilization of financial services. The term "Islamic fintech" is employed to describe the transformation of financial services and processes through the integration of technological innovations, including blockchain, artificial intelligence, and big data, with the principles of Islamic finance. The advent of Islamic fintech

DOI: 10.4018/979-8-3693-8079-6.ch006

has precipitated a notable surge in demand for Islamic financial products across regions such as the Middle East, Southeast Asia, and Africa. This study assesses the compatibility of fintech with the ethical and operational framework of Islamic finance, with particular emphasis on interest-free banking, risk sharing, and transparency. Furthermore, the study evaluates the potential benefits and obstacles associated with the digitization of the Islamic finance sector, including the sustainable growth and the advancement of innovative solutions. Additionally, the study assesses the impact of regulations and policies in this context.

1. INTRODUCTION

The rapid advancement of technology has had a profound impact on numerous facets of human existence, including the financial sector, which plays a pivotal role in our lives. The digitalization and technological innovations have transformed the provision of financial services, accelerating the sector and creating new opportunities. In addition, markets have undergone significant changes as a result of these technological developments and digital transformations, entering a state of constant change and renewal. A review of developments over the past five years reveals that financial technology (fintech) has had a profound impact on the provision of financial services, fundamentally altering traditional banking and finance processes. Fintech has not only facilitated the transition of conventional financial services to digital platforms but has also spearheaded the advent of novel business models in domains such as payment systems, digital wallets, blockchain, and AI-driven financial solutions. The advent of fintech has precipitated a paradigm shift in the financial sector, affecting domains as diverse as banking, investment, payments, insurance, and other financial systems. The innovative solutions offered by fintechs have enabled users to access faster, more economical, and more accessible financial services, while also gradually increasing consumer financial literacy. By meeting societal demands for more efficient and inclusive financial services, as well as increasing financial mobility and complexity, fintech has contributed to the emergence of a "new" financial order, replacing the distrust of the traditional financial system on a global scale.

The crises experienced in the past and the inadequacy of the traditional financial system, along with the necessity of sustainable economic growth and development on a global scale, have become a priority in today's financial world. In this context, the development of novel financial models that facilitate sustainable finance is of paramount importance. In recent years, the Islamic financial system has also witnessed a surge in interest from global financial markets. The rapid developments in the field of fintech have accelerated the digitalization process of Islamic finance, and these technological innovations have paved the way for radical changes in the world of Islamic finance. Consequently, a novel ecosystem designated as Islamic fintech has emerged, and Islamic financial markets have persisted in their expansion by capitalizing on the digital transformation.

The integration of products and services based on Islamic finance principles with fintech has provided an important opportunity for the global expansion of participation banking and Islamic finance. As a consequence of the integration of Islamic finance with fintech, a sub-branch of Islamic fintech has emerged, with the aim of offering fintech innovations that comply with Islamic law and jurisprudence. This is achieved by providing financial solutions based on the principles of interest-free banking, risk sharing, transparency, and ethical finance. The principal objective of this study is to assess the present role and prospective scope of Islamic fintech, with particular consideration of the novel technologies that are being employed. Furthermore, this study aims to provide a framework for future research on Islamic fintech and to facilitate discourse on the prospective evolution of fintech by offering predictions. In the study, an examination will be undertaken of the digitalization process of Islamic financial services, with a focus on the new business models offered by fintech in the context of emerging technologies such as blockchain, artificial intelligence, big data, and smart contracts, as well as the impact of digitalization on Islamic financial products. This study will analyze the current status and development trends of Islamic fintech on a global scale. Furthermore, the manner in which Islamic fintech initiatives respond to consumer needs and the role these technologies play in terms of transparency, trust, and ethical finance principles will also be a significant area of focus. Finally, the study will conclude with an examination of the prospective potential of Islamic fintech and its capacity to extend the reach of the Islamic finance system. The opportunities and challenges of digitalization in the Islamic finance sector will be evaluated in terms of the sector's sustainable growth and the development of innovative solutions. Additionally, the role of regulations and policies in this area will be examined.

In the literature review section of the study, previous studies on the integration of fintech solutions into the Islamic finance ecosystem were examined and presented in order to identify gaps in the existing literature on this topic. The extant literature demonstrates that Islamic fintech has considerable potential for providing financial services that are in accordance with Islamic finance regulations. Nevertheless, several challenges remain, particularly with regard to regulatory compliance, technological adaptation, and the effective implementation of Shariah compliance procedures on a global scale. The third section of the study addresses the emergence of the Islamic fintech ecosystem, which has resulted from the integration of Islamic finance with new technologies. It also considers the future directions and potential ethical implications of this ecosystem. Islamic fintech is a rapidly expanding field that integrates cutting-edge technologies in accordance with Islamic finance principles. The integration of advanced technologies, including blockchain, artificial intelligence, big data, and smart contracts, is facilitating the acceleration, transparency, and accessibility of Islamic financial services. This digital transformation meets the needs of not only Muslim communities but also a wide range of customers seeking interest-free and ethical financial services on a global scale. By facilitating greater financial inclusion, Islamic fintech provides alternative solutions for individuals and small and medium-sized enterprises (SMEs) that are currently unbanked, while also introducing a novel and sustainable perspective to the traditional financial system.

2. LITERATURE REVIEW

There is a substantial body of research examining the influence and expansion of fintech on a global scale, including in Turkey. However, there is a notable increase in the number of studies focusing on Islamic fintech, yet the literature remains limited in scope and depth. El-Gamal (2006), one of the important studies in this field, discussed the historical development of Islamic finance and the main problems encountered in this process, and presented a comprehensive framework on how Islamic finance can be integrated with modern financial technologies. In their studies, Karim and Archer (2013) examine the challenges faced by Islamic financial institutions in the digitalization process and the opportunities that emerge from it. In its 2019 global fintech report, Gateway (2019) examined the impact of Islamic fintech on the global economy. It was found that this report emphasizes the rapid growth of Islamic fintech, particularly in regions such as the Middle East, Southeast Asia, and Africa. Alam et al. (2019) examined the potential of Islamic fintech and the innovations made in this field, as well as the ways in which Islamic finance is transformed by fintech. In a discussion of the regional distribution of Islamic fintech and its contribution to economic development, DinarStandard (2018) highlighted the importance of this sector in fostering sustainable growth. Alblooshi, Head of the Innovation Center and Fintech at Dubai International Financial Centre (DIFC), discussed the contribution of Islamic fintech to sustainable development, emphasizing the potential of Islamic fintech to develop sustainable financial solutions that support economic development. In a study published in 2019, Pietro Biancone and colleagues evaluated Islamic fintech initiatives in terms of innovation and entrepreneurship. They analyzed the potential for developing innovative business models in the field of Islamic fintech and the risks these initiatives may face. Furthermore, Oseni and Ali (2019) investigated the evolution of the Islamic fintech ecosystem and optimal practices within this domain, offering theoretical and practical insights into Islamic fintech and comprehensive guidance on best practices and strategies. Hassan et al. (2022) reached the conclusion that the Islamic finance sector is experiencing unprecedented growth as a result of investments in fintech. The two areas that have been identified as experiencing the highest rates of growth within Islamic finance are blockchain and crowdfunding. The article provides an advanced overview of blockchain and crowdfunding.

In their 2019 study, Abojeib and Habib investigated the potential applications of blockchain technology in the management of Islamic social responsibility institutions. The authors posit that this technology provides transparency, trust, and traceability, thereby enhancing the efficacy and reliability of donation processes such as zakat and sadaqah. The study illuminates the potential of blockchain technology in this domain, demonstrating its capacity to enhance the management of Islamic social responsibility initiatives. Rabbani et al. (2020) investigated the influence of blockchain technology on Islamic financial services and its capacity to fulfill the transparency and reliability requirements of Islamic finance. They examine the potential benefits of blockchain technology for Islamic finance and the challenges this technology may face in practice.

The global digitalization process of Islamic finance has progressed at disparate rates across different regions. While these studies address the integration of Islamic finance with fintech from a global perspective, it is notable that there have also been significant developments in this field in Turkey. Indeed, the development of Islamic fintech in Turkey has gained momentum in line with global trends. The academic literature on Islamic fintech initiatives in Turkey has concentrated on examining the potential and development process of this field. In a comprehensive analysis of the development of the Islamic fintech ecosystem in Turkey and innovations in this field, Turan (2021) provides a valuable contribution to the existing literature. Turan's analysis encompasses the digitalization process of Islamic finance in Turkey, the contributions of fintech initiatives to financial inclusion, and the legal obstacles encountered. Aktürk (2021) provides an analysis of the integration processes of Islamic banks with fintech in Turkey, identifying the challenges inherent in this process and examining how banks protect Islamic principles in the digitalization process. In his study examining the integration of Islamic financial institutions with blockchain technology in Turkey, Demirdöğen (2023) discussed the advantages and difficulties that this technology will provide to Islamic financial institutions. Concurrently, in his study evaluating the legal infrastructure of Islamic fintech initiatives in Turkey and the regulations pertaining to this field, he examines the prospective effects of Islamic fintech on the Turkish economy. Blockchain technology is a pivotal element in the advancement of Islamic fintech. Concurrently, blockchain technology is at the core of Islamic fintech initiatives in Turkey. Samar et al. (2020) examined the potential of blockchain-based Islamic fintech solutions in Turkey and the manner in which these solutions can be integrated into Islamic financial institutions. They addresses the question of how blockchain technology can be aligned with the fundamental principles of Islamic finance and the potential applications of this technology in Turkey.

3. ISLAMIC FINTECH: THE INTEGRATION OF INNOVATIVE TECHNOLOGIES WITH ISLAMIC FINANCE

Islamic fintech is a rapidly growing industry that aims to provide financial services in accordance with Islamic principles while leveraging the power of technology to make financial transactions more efficient, convenient, and accessible. It advances financial inclusion and social responsibility by addressing the distinctive financial requirements of consumers through cutting-edge technologies. Consequently, the future of Islamic fintech is influenced by the accelerated uptake of technological advancements and the formulation of regulations in this domain (Finterra, 2024).

While there is no distinction in technology between traditional and Islamic fintech, there are notable differences in how these technologies are applied, operated, and how they align with Islamic principles, regulations, and ethical standards. Consumer needs also reflect these differences. The following is a detailed overview of the aforementioned differences:

- Compliance with Islamic Principles: Islamic fintech companies must develop products and services in accordance with Islamic finance principles and make them available to consumers. It is therefore essential that financial transactions conducted through Islamic fintech initiatives are not based on interest, do not contain uncertainty (gharar), and are conducted in an ethical and fair manner.
- Compliance with Ethical Principles: Islamic fintech companies do not engage in activities that are detrimental to society or the environment, such as gambling and alcohol, which are not aligned with Islamic principles. In the same way that Islamic finance is based on moral responsibility, these companies must also take this approach.
- Compliance with Participation and Risk Principle: Islamic finance principles encourage risk sharing, partnership, and profit sharing, as opposed to risk transfer. In accordance with Islamic principles, those who offer and request financing must assume this risk in a collaborative manner. Consequently, Islamic fintechs must develop their business models in accordance with this foundation.
- Compliance with Regulatory Principles: Islamic fintech companies are also required to adhere to the regulations that apply to Islamic financial institutions. It is imperative that they comply with the regulations set

forth by the authorities that oversee compliance with Islamic principles and regulations.

Islamic financial institutions and technology companies are actively engaged in the provision of Islamic digital financial services. The preparation of technology infrastructure for the development of Islamic financial institutions and the Islamic digital ecosystem, along with investment in online and mobile banking channels, compliance with regulations, and support for the digital environment, is gradually gaining momentum as a means of transforming Islamic finance with Islamic fintechs (Hud Saleh Huddin, Lee, and Mansor, 2022). As illustrated in Figure 1, the foundational elements of Islamic fintech are explored within four principal categories: digitalization strategic issues, transparency, accessibility through digital channels, and technology.

Figure 1. The operational framework of Islamic fintech (Adapted from Fintech ve İslami Finans; Dijitalleşme, Kalkınma ve Yenilikçi Yıkım by Alam et al., 2021)

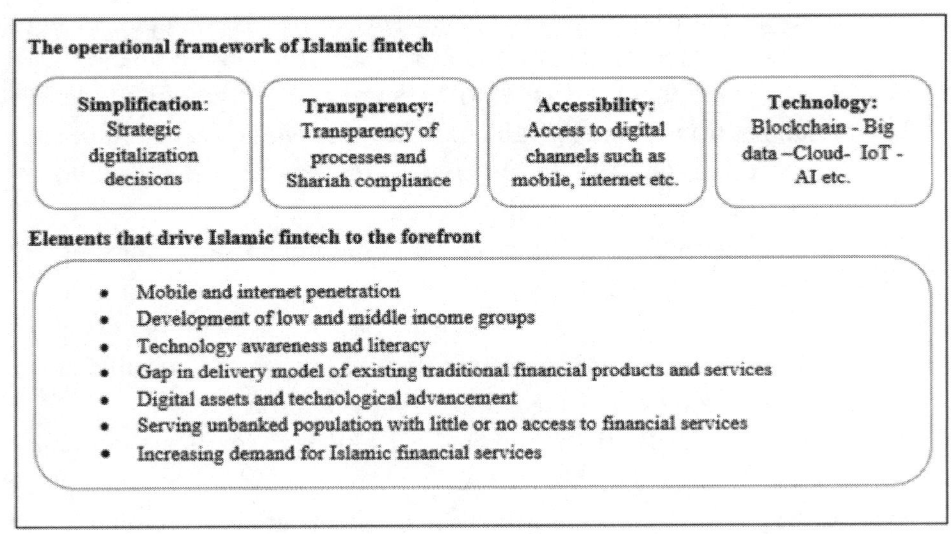

In the field of Islamic fintech, a multitude of financial solutions are developed through the application of novel technologies. As a foundational element of Islamic fintech, a range of technologies are employed, including blockchain, cloud computing, big data analytics, robo advisors, IoT (Internet of Things), machine learning, and artificial intelligence. These technological infrastructures enhance transparency in Islamic financial services, guarantee the traceability of processes, facilitate data-driven decision-making, and bolster the reliability of financial products. Blockchain technology, in particular, facilitates the Sharia-compliant operation of Islamic finance, which is based on the principles of transparency and trust. It also expedites processes through the use of smart contracts and reduces costs. In contrast, data analytics enables the provision of bespoke financial solutions through the analysis of customer behaviour, thereby facilitating access to a broad customer base. Customer decision support systems enable the provision of tailored financial recommendations based on an analysis of customer data, including past purchasing patterns and product preferences. Consequently, customers are furnished with the tools necessary to make informed financial decisions, thereby enhancing customer satisfaction. Nevertheless, another significant technological solution offered by Islamic fintech is the analysis of fraudulent activity. In order to mitigate the risk of fraud, machine learning and artificial intelligence algorithms are employed to monitor transactions in real-time and identify anomalous behavior. Consequently, customer security is enhanced, and financial transactions are conducted in a secure environment. Such technologies enhance the reliability and transparency of Islamic finance, thereby reinforcing trust in the sector.

In examining financial services and the fintech ecosystem in accordance with these principles, it becomes evident that the primary distinction between existing services and emerging technologies is the evolving consumer base and their needs. Consequently, there is an increasing demand for ethical financial services, particularly interest-free options, from both Muslim and non-Muslim individuals. This has resulted in a notable increase in interest in Islamic fintech solutions. This growing demand for ethical financial products and services that align with the values of both Muslim and non-Muslim consumers demonstrates that Islamic fintech solutions have a broader appeal than traditional fintech offerings. They are positioned for substantial global expansion. Islamic fintech companies employ cutting-edge technologies to provide consumers with convenient, cost-effective, user-friendly, and innovative technology-focused solutions that adhere to ethical standards. These initiatives employ technological advances in their business models to serve users in areas such as payments, money transfers, investments, asset management, crowdfunding, investment consultancy, insurance, and financing.

The most sophisticated technologies currently employed in the provision of these services include blockchain technology, cloud computing, big data analytics, robo-advisor, the Internet of Things, machine learning, and artificial intelligence.

In the aftermath of the 2008 global financial crisis, blockchain technology has emerged as a pivotal area of focus within the financial ecosystem. Initially employed for investment purposes, cryptocurrencies have subsequently been adopted as digital payment instruments, thereby further enhancing the technology's appeal. Blockchain provides a decentralized platform for the transparent recording of transactions, thereby ensuring the traceability and integrity of financial activities and thus gaining consumer trust. While banks are regarded as dependable intermediaries in conventional financial systems, the slowness, physicality, and mounting transaction costs of these systems have prompted the pursuit of more innovative solutions. At this juncture, blockchain technology is attracting significant interest within the financial sector, offering a novel and promising perspective (Alam et al., 2021).

The structure of blockchain technology, which aligns with the principles of Islamic finance, including transparency, security, and the prohibition of gharar (uncertainty), also positions this technology as a significant contributor to the field of Islamic finance. In particular, the transparency and security advantages provided by smart contracts facilitate the acceptance of Islamic financial products in the global market. Islamic fintech initiatives employ blockchain technology to facilitate processes such as sukuk issuance, crowdfunding, digital payment systems, and to implement alternative insurance models (takaful) for mutual aid purposes (Hacak & Gürbüz, 2019). Indeed, as of 2023, the global market size of blockchain technology in the financial sector has exceeded 10 billion dollars, with applications of this technology in the field of Islamic finance continuing to develop rapidly (Grand View Research, 2023).

Table 1. Blockchain-based Islamic fintech projects (Kılıç & Türkan, 2023)

Financial Institution	Country	Blockchain Initiative/Use-case
Al Rajhi Bank	Saudi Arabia	Services for international money transfers utilising blockchain technology.
Abu Dhabi Islamic Bank (ADIB)	UAE	Utilise blockchain for trade finance and intelligent contracts.
Islamic Devel-opment Bank (IsDB)	Multinational	Partnered with entrepreneurs to investigate the potential of blockchain for Islamic finance solutions.
Kuwait Finance House (KFH)	Kuwait	Ripple's blockchain-based solution for international transactions was evaluated.
Emirates Islamic Bank	UAE	Using blockchain technology to improve finance ser-vices and security.
Finterra	Singapore	Provides a blockchain-based platform for adminis-tering end-to-end Islamic financial services, including waqf (charitable trust) and crowdfunding.

The potential applications of blockchain technology in Islamic finance are the subject of an increasing amount of research. The increasing prevalence of cryptocurrencies and blockchain technology is poised to yield substantial implications for Islamic banking, which has ushered in a paradigm shift in the global financial landscape. While the growing interest in cryptocurrencies presents certain challenges in the context of Islamic finance, it also offers a range of promising opportunities. The decentralized and digital nature of cryptocurrencies has the potential to facilitate financial inclusion to the extent that they are organized in accordance with Islamic finance principles. Blockchain technology, which is based on the principles of transparency, reliability, and justice as defined by Islamic finance, has the potential to transform the field of Islamic fintech. This technology, particularly due to its immutable record-keeping functionality, facilitates the collection and distribution processes of zakat, thereby strengthening the trust of donors. Moreover, blockchain technology provides the potential for increased efficiency through the automation of smart contracts for Islamic real estate investments, sukuk, and other Islamic financial transactions. Blockchain technology, which also demonstrates the potential for cost reduction and increased speed in cross-border payments, makes an innovative contribution to the Islamic finance ecosystem by facilitating the tokenization of tangible assets (Selcuk & Kaya, 2021). Table 1 presents an overview of select Islamic Fintech initiatives employing blockchain technology.

The implementation of cloud computing technology has resulted in a substantial transformation in numerous industries, particularly within the financial sector, in recent years. The provision of virtual infrastructures accessible via the internet, as opposed to physical servers and hardware, greatly facilitates the processes of storing, processing, and analyzing data. The flexible structure, cost advantages, and scalability of cloud computing enhance operational efficiency in companies while facilitating faster and more secure services than traditional systems. In particular, Islamic financial institutions provide enhanced traceability in alignment with the tenet of secure data storage and transparency through this technology. Moreover, the advent of cloud computing has enabled fintech initiatives to offer a range of services, including digital payment systems, crowdfunding, and takaful, in a more efficient manner. This has contributed to the growth of Islamic finance on a global scale (Alam et al., 2021).

Businesses are seeking to facilitate smooth customer service while simultaneously enhancing operational efficiency. It is the desire of customers to receive a service that is straightforward, rapid, and readily accessible from start to finish. The financial sector is responding to the needs of businesses and customers by offering innovative solutions that leverage big data analytics and Internet of Things (IoT) technologies. The application of big data analytics enables financial institutions to make more informed decisions by facilitating the analysis of large and complex data sets that cannot be processed with traditional methods. This technology offers significant advantages to financial service providers in areas such as customer behavior analysis, risk management, and personalized product development. The Internet of Things (IoT) enables physical devices to exchange data with each other over the internet, thereby facilitating the implementation of applications such as real-time monitoring, data collection, and process automation in the financial sector. In particular, Islamic financial institutions integrate these technologies with the objective of minimizing risks, increasing transparency, and providing more effective services to customers. This is achieved through the use of big data and IoT technologies. These innovations facilitate the acceleration of the digitalization of Islamic finance, thereby conferring a global competitive advantage.

Robo-advisors represent a novel approach within the financial technology sector. In the domain of Islamic fintech, it serves as a pivotal instrument for digital transformation. Robo advisors, which are more cost-effective and accessible than traditional financial advisory services, distinguish themselves by their capacity to provide personalized investment strategies in alignment with Islamic finance principles. These systems offer algorithmic solutions that are tailored to the financial objectives, risk tolerance, and investment preferences of the user in accordance with Islamic principles, thereby facilitating the creation of interest-free, ethical, and transparent investment strategies. In the context of Islamic fintech, robo advisors facilitate the automation of processes such as asset management, portfolio optimization, and risk management in accordance with Islamic rules, thereby enhancing the accessibility of Islamic financial products to a broader demographic. Consequently, robo advisors facilitate the digitalization of Islamic finance and bolster the advancement of Islamic fintech initiatives, which are a rapidly expanding domain (Alam et al., 2021).

The advent of machine learning and artificial intelligence (AI) has opened the door to a new era of innovative applications in Islamic fintech, many of which were previously not feasible. While machine learning provides more accurate insights into customers' financial preferences through self-improving algorithms in data analysis, AI goes beyond traditional processes and offers intelligent, automated financial solutions. In the context of Islamic banking, AI-based systems are capable of conducting rapid and transparent audits in accordance with Islamic finance regulations. Additionally, machine learning can be employed to generate personalized interest-free investment and financing recommendations based on an individual's financial history and risk profile. These technologies are employed in Islamic fintech initiatives with the objective of enhancing service quality in a number of areas, including the automation of smart contracts, the optimization of crowdfunding platforms, and the personalization of takaful (Islamic insurance) solutions. The application of artificial intelligence and machine learning in customer service, foreign exchange and halal stock exchange transactions, asset management, security and the prevention of financial crimes, and the enhancement of credit scoring processes contribute to the strengthening of Islamic finance in the context of global competition, the development of an innovative and sustainable financial ecosystem.

Table 2. Top 20 Islamic fintech startups (Gateway, 2024 - redesigned by the author)

Rank	Company Name	Country	Score	Thecnology Category
1	Wahed Invest	United States	4.5	Robo Advisors, Crowdfunding
2	Tabby	Saudi Arabia	4.3	Alternative Finance
3	Tamara	Saudi Arabia	4.3	Alternative Finance
4	Hijra (Alami Sharia)	Indonesia	4.2	P2P Finance
5	Kitabisa	Indonesia	4.1	Crowdfunding
6	Lendo	Saudi Arabia	3.9	P2P Finance
7	Alif	United Kingdom	3.7	Challenger Banks
8	PolicyStreet	Malaysia	3.5	TakaTech
9	Ethis	Malaysia	3.5	P2P Finance
10	Musaffa	United States	3.4	Trading & Investment
11	Fasset	UAE	3.3	Blockchain & Cryptocurrencies
12	Ayan Capital	United Kingdom	3.3	Challenger Banks
13	Funding Souq	UAE	3.3	P2P Finance
14	Beehive	UAE	3.3	P2P Finance
15	Launchgood	United States	3.2	Crowdfunding
16	Manzil	Canada	3.1	Alternative Finance
17	Algbra	United Kingdom	3.1	Alternative Finance
18	Iman	Uzbekistan	3.1	Trading & Investment
19	Takadao	Saudi Arabia	3	Blockchain & Cryptocurrencies
20	Madfu	Saudi Arabia	3	Alternative Finance

As indicated in the "Top 30 Digital Islamic Economy Startups" report prepared by Salaam Gateway, financial technology startups exert a considerable influence within the Islamic finance ecosystem (Salaam Gateway, 2024). The top 20 of these startups are shared in Table 2. Digital investment and financing platforms, including Wahed Invest, Ethis, and IFIN, are at the vanguard of the digital transformation of Islamic finance. These fintech startups reinforce their leadership roles in the sector by appealing to all consumers without discrimination and to a broader audience seeking financial solutions that are not adequately addressed by the existing financial system.

An analysis of the activities of the top 30 digital Islamic economy startups and the development of Islamic fintech reveals that they encounter a multitude of challenges pertaining to compliance with both Islamic finance rules and legal regulations. One such challenge is that of auditing. The digitization of Islamic financial products and their integration with fintech solutions necessitates a meticulous auditing process to ensure compliance with Islamic finance principles. For instance, Dubai Islamic Bank collaborates closely with the Shariah Advisory Board to ensure that its fintech solutions are aligned with Islamic finance principles. Such collaborations facilitate the ethical and transparent growth of Islamic fintech, while also ensuring the provision of reliable financial services to customers (Noronha, 2020). Furthermore, the regulatory challenges currently faced by Islamic fintech on a global scale have yet to be addressed. While some countries have enacted promising regulations in this domain, a comprehensive legal framework for Islamic fintech at the global level remains to be fully established. It is therefore of great importance to strengthen cooperation between fintech startups, Islamic financial institutions, and regulatory authorities in order to establish industry standards in the Islamic fintech ecosystem. A regulatory structure that is compatible with technological developments and supports innovation will facilitate the growth of Islamic fintech on a global scale, enabling it to meet the evolving financial needs of Muslim consumers worldwide (Faster Capital, 2024).

4. FUTURE DIRECTIONS AND POTENTIAL IMPACTS

Islamic finance has undergone a significant transformation from a structure that exclusively caters to the needs of Muslims to a new financial system and ecosystem that is rapidly expanding in the global financial market and reaching double-digit growth rates. Initially confined to Muslim countries and communities, this financial model has since expanded to a universal scope, operating globally and competing with traditional solutions as an alternative financial and investment tool. It has become a significant sector in its own right (Alam et al., 2021). Additionally, Islamic fintech companies are at the vanguard of the digitalization of conventional Islamic financial services. This transformation process entails the implementation of an array of digital tools and cutting-edge technologies, with the objective of expediting, enhancing transparency, and facilitating greater accessibility to Islamic financial services. Figure 2 presents the Islamic Fintech Landscape in 2024.

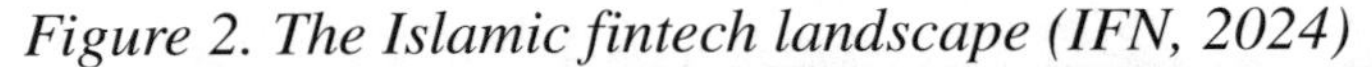
Figure 2. The Islamic fintech landscape (IFN, 2024)

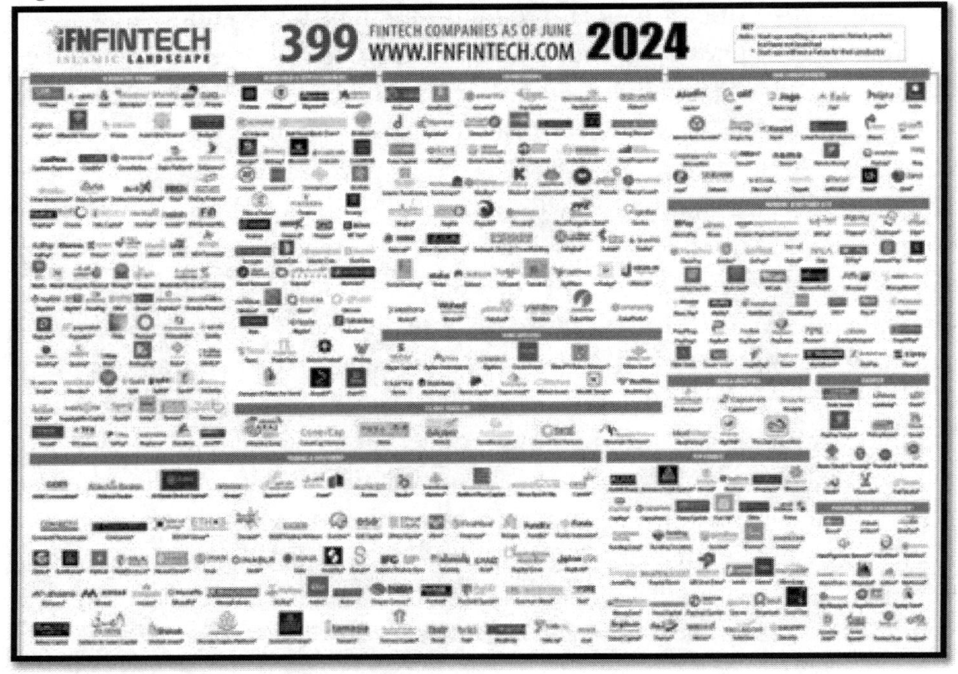

The application of digital technologies by Islamic fintech facilitates the adaptation of traditional Islamic financial services to the requirements of the contemporary era, while simultaneously providing innovative solutions. These solutions address the needs of both the Muslim community and a broader customer base seeking services in alignment with Islamic finance principles. In 2023, the global Islamic finance market exceeded $3 trillion, with the market share of Islamic fintech in this area rapidly increasing (Greggwirth, 2022). Concurrently, the global fintech market surpassed $500 billion by 2023 and is projected to reach $1 trillion by 2028 (Statista, 2024). As of 2024, there were more than 350 Islamic fintech startups worldwide, 35% of which provided payment services and 25% provided financial inclusion and micro-finance services (IFN Fintech News, 2024). Despite their relatively modest scale, Islamic fintech initiatives operating in the field of Islamic finance are experiencing rapid growth and development (Orhan, 2023).

A review of the data presented in the "2024 Global Islamic Fintech Report and IFN Fintech Landscape" reveals a noteworthy expansion in the Islamic fintech sector on the global scale. This trend is particularly pronounced in

Malaysia, Indonesia, the United Kingdom, and several Gulf states. As indicated in these reports, there are 147 Islamic fintech companies currently operational in Asia, 107 in the Gulf countries, and 77 in Europe. This ecosystem, which encompasses a total of 417 Islamic fintech companies operating on a global scale, is notably concentrated in domains such as digital banking, payment systems, and asset management. The market size of the Islamic fintech sector is estimated to have reached $138 billion in 2022, representing a compound annual growth rate (CAGR) of 17.3%. It is projected to reach $306 billion by 2027. This growth rate is considerably higher than that observed in the traditional fintech sector, which recorded a 12.3% growth rate. In terms of regional distribution, countries such as Malaysia, Saudi Arabia, the United Arab Emirates (UAE), Indonesia, and Iran collectively account for approximately 85% of the global Islamic fintech market. (Salaam Gateway, 2024). The figure 3 presents the estimated sizes of the Islamic fintech markets in various countries in 2024, expressed in billion U.S. dollars.

Figure 3. Islamic fintech market sizes by country 2024 (Gateway, 2024)

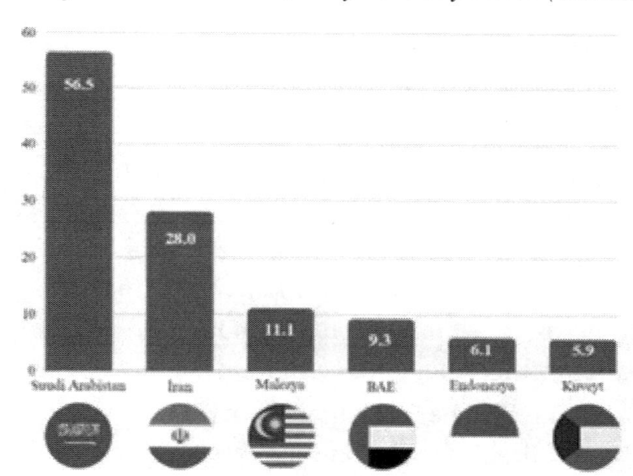

In Europe, developments in the Islamic fintech field, spearheaded by the United Kingdom, are attracting considerable attention. While London is becoming a hub for Islamic finance and fintech, interest in Islamic finance-compatible fintech solutions is also on the rise in other European countries, including France and Germany. This interest is predicated on the growth of the Muslim population in Europe and the concomitant demand for ethical financial services. An analysis of the Global Islamic Fintech Report highlights the pivotal role of emerging technologies, including blockchain, artificial intelligence, and big data, in the development of Islamic fintech solutions. These technologies offer innovative solutions that are compatible with the principles of Islamic finance, particularly in regard to transparency, speed, and security. To illustrate, blockchain technology enhances the traceability of financial instruments such as Sukuk, thereby upholding the core tenets of Islamic finance, namely transparency and risk-sharing.

The global reach and adoption of Islamic FinTech applications is rapidly expanding. These FinTech applications, which offer solutions in accordance with Islamic finance principles, have gained significant momentum in countries such as Malaysia, Indonesia, England, and Turkey, especially in the Gulf countries. In particular, in Gulf countries where Islamic finance is widely practiced, financial technology solutions are experiencing a rapid uptake and are attracting a diverse range of users. Southeast Asian countries such as Malaysia and Indonesia are also at the forefront of Islamic FinTech, developing their digital infrastructures and increasing their innovative solutions in the field of Islamic finance.

Turkey has made a notable impact on the Islamic FinTech landscape in recent years, with substantial investments in the sector. Turkey, with a long history in Islamic banking, is rapidly developing its Islamic FinTech sector thanks to new regulations on digital banking and supportive government policies. The launch of new digital participation banks in Turkey demonstrates the significant growth potential of the Islamic FinTech ecosystem. It is anticipated that Islamic FinTech applications in Turkey will become widely adopted across various sectors, including payment systems, digital participation banking, individual financing, and Islamic investment platforms.

There is a growing number of Islamic FinTech applications worldwide, some of which are attracting international attention due to the innovative solutions they offer. For instance, platforms such as Souqalmal from the Gulf countries, Ethis Ventures based in Malaysia, and Yielders based in the UK operate in areas such as investment, lending, savings, and insurance in accordance with Islamic finance principles (Cengiz and Ozkan, 2023). In Turkey, local initiatives such as Insha Ventures participate in the Islamic FinTech ecosystem and bring Islamic finance solutions to users in digital environments. Table 3 provides an overview of prominent examples from the world and Turkey.

Table 3. Prominent initiatives of Islamic fintech in the world and in Turkey (Cengiz and Ozkan, 2023; Akturk, 2024)

Islamic FinTech Initiative	Initiative's Scope of Operations and Key Features
Ethis Türkiye	It provides entrepreneurs with the opportunity to raise funds to implement their projects and offers investors investment options that are in line with their religious sensitivities. Ethis Türkeye provides the opportunity to support projects based on Islamic finance principles and offers innovative financial models that meet the needs of both entrepreneurs and investors.
EthisCrowd	EthisCrowd, the first Islamic crowdfunding platform launched in Singapore, is a real estate Islamic FinTech initiative that places great emphasis on affordable housing allocation and focuses on social responsibility projects.
Faizsiz Teklif	It is a financing platform structured in accordance with Islamic finance principles. It creates a different alternative for potential customers seeking interest-free financing by offering interest-free financial institutions' credit and financing solutions on a single platform for financing needs such as housing, vehicles, needs, etc. or financial products such as insurance policies.
Goldframer	It is an Istanbul-based Islamic fintech platform that allows for partial investment in Turkish-Islamic artworks through crowdfunding and aims to facilitate art investments in a digital environment. Launched in 2024, Goldframer offers investors the opportunity to make equity-based investments in Islamic artworks, ensuring both the protection of these valuable cultural assets and the benefit of their financial return potential.
Helal Launcher	Founded in the UK as a halal and multi-party crowdfunding platform, the company invests in halal projects and ideas.
Hijra	Hijra, known as ALAMI Group, is an Islamic Fintech company that provides financial solutions for individuals and businesses, including a peer-to-peer lending platform, mobile banking app, business accelerator and research institute, all in accordance with Islamic finance principles.
Islamic Finance Guru	Founded in 2015 and headquartered in the UK, the company provides halal investment, Islamic mortgage and financing services to individual and corporate clients in need. It also offers all types of syndicated transactions/investments for Sharia-compliant start-ups.
Kapital Boots	Based in Singapore and based on partnership methods, Capital Boost is an Islamic crowdfunding platform established to provide support to SMEs in need of financing to grow their existing businesses.
Kestrl	This digital banking company operates through the Bank of England and offers interest-free payment and wealth management services in accordance with ethical values.
LaunchGood	It is a crowdfunding platform that helps entrepreneurs and projects meet their financing needs. Founded in 2013, this platform is an Islamic FinTech company that aims to provide support on a global scale, focusing especially on projects that create social impact and provide social change.

continued on following page

Table 3. Continued

Islamic FinTech Initiative	Initiative's Scope of Operations and Key Features
Manzil	As the first Canadian-based Islamic Fintech company, Manzil stands out as the first and only full-service provider of Sharia-compliant housing finance in Canada.
Marhaba DeFi	It is a digital platform that provides decentralized finance (DeFi) services in accordance with the principles of Islamic finance. Founded in 2020, Marhaba DeFi provides its users with the opportunity to conduct financial transactions on crypto assets in accordance with Islamic rules, adhering to the ethical and transparency principles of Islamic finance.
Menapay	The first Turkish FinTech startup to receive investment from CoinShares Ventures, this company offers money transfer and payments, crypto buying/ selling, token wallet, token storage and management services in Turkey.
QardHasan	Founded in 2015 to support students online, this system offers interest-free student loans for Muslim students.
Rizq	It is a UK-based Islamic fintech startup and is positioned as the country's first digital banking app that is fully compliant with Islamic finance principles. Founded in 2020, Rizq was developed specifically to meet the needs of the Muslim community in the UK.
Scola Fund	Scola Fund, a platform first launched in Malaysia to provide scholarship support to students in need, is an Islamic FinTech initiative that has established a network in 30 countries.
Tabby	Tabby is the first unicorn and licensed Islamic fintech company to offer a Buy Now, Pay Later solution operating in the MENA region. Founded in 2019, the company offers a "Buy Now, Pay Later" (BNPL) model to provide consumers with flexibility in their shopping.
Wahed Invest	Wahed Inc. (Wahed Wahed Invest) is a global digital investment platform that provides investment services in accordance with ethical and value-oriented Islamic principles. Founded in 2015, Wahed is known as the company that launched the world's first fully Sharia-compliant digital investment platform.
Wigorta	It is an insurance platform designed in accordance with Islamic insurance principles. Unlike traditional insurance systems, this initiative offers insurance products that do not include interest and risk sharing. The policies offered by Wigorta are structured in accordance with Islamic finance rules and provide a fair and transparent insurance experience to its users.
Yielders	Originally founded in the UK, Yielders' scope of activity and practice is the pioneer of an innovative entrepreneurial sector that includes the principles of real estate acquisition and rental income sharing.

A review of the 2024 data illuminates the regional diversity of the Islamic fintech ecosystem and its potential for growth on a global scale. Malaysia, Indonesia, Saudi Arabia, and the UAE are at the vanguard of the sector, with fintech solutions from these countries spearheading the digitalization of Islamic financial services on a global scale. This growth is anticipated to be bolstered by further investment and innovation in the forthcoming years. The data indicates that Islamic fintech has the potential to appeal to a broader audience in this field by combining modern technology with Islamic finance, thereby strengthening the role of Islamic finance in the global economy and increasing its share of the financial services market. The potential of Islamic fintech extends beyond the digitalization of financial services. It also presents opportunities to enhance financial inclusion, facilitate access to new markets, and develop novel products and services in alignment with the tenets of Islamic finance. In this context, Islamic fintech is poised to assume a pivotal role in the global financial ecosystem, becoming a pivotal factor in the transformation and evolution of financial services in the future (Gateway, 2024).

Gateway (2024) has presented the detailed overview of the regional and sectoral classification of Islamic Fintech startups presented in Table 4. It demonstrates that alternative finance represents the leading sector, with payments, fundraising, and lending following closely behind. By region, the most advanced areas are Southeast Asia, the Middle East and North Africa (MENA) region and Europe. The situation in other regions demonstrates that global gaps persist, with limited activity observed across the board. Insurance and capital markets are the least developed sectors globally. While digital assets, wallets, web3, and decentralized finance categories are increasing in popularity, there are significant opportunities for further growth and widespread expansion due to the use of blockchain, smart contracts, and artificial intelligence (AI) technologies.

Table 4. Number of Islamic fintech startups by region (Gateway, 2024, The table was adapted from the table presented by Akturk, 2024).

Region/Category	Southeast Asia	MENA GCC*	Europe	South & Central Asia	North America	Sub-Saharan Africa	MENA Other	TOTAL
Alternative Finance	24	21	12	3	1	2	1	64
Interface Applications	4	6	3	1	2	0	0	15
Capital Markets	3	5	1	0	0	0	0	9

continued on following page

Table 4. Continued

Region/Category	Southeast Asia	MENA GCC*	Europe	South & Central Asia	North America	Sub-Saharan Africa	MENA Other	TOTAL
Deposits and Financing	11	11	14	3	2	5	2	48
Digital Assets	3	7	1	0	4	0	0	27
Facilitating Technologies	12	10	1	0	1	0	1	25
Insurance	8	4	1	0	1	1	0	15
Payments	18	31	5	3	10	1	3	63
Fund Raising	31	9	11	4	2	2	2	61
Social Finance	5	1	3	1	4	1	1	16
Technology Providers	5	7	4	5	1	2	0	19
Asset Management	10	19	7	3	10	2	1	55
Total	138	131	66	31	26	15	10	417

*MENA-GCC: Middle East and North Africa Gulf Cooperation Countries

A review of fintech applications, developments, and studies in Turkey reveals that its central location connecting Asia and Europe makes it a financially important country. Consequently, Turkey plays an instrumental role in the advancement of financial and digital technologies on a global scale. Although the concept of fintech was first used in Turkey in 2001, it has not remained a fintech concept over the years. Instead, it has become the ecosystem in which fintech companies operate, with the influence of the state and other ecosystem supporters. Following the year 2020, the payment services sector was identified as the primary area of operation for fintech companies, a trend observed globally and in Turkey. This was followed by cryptocurrency applications. The advancement of fintech in Turkey gained significant traction following the year 2016. The Finance Office of the Presidency of the Republic of Turkey has initiated a series of significant studies in the domain of financial technologies. Additionally, it has devised a range of projects and policies with the objective of reinforcing Turkey's strategic standing in this field. As part of these studies, Islamic fintech emerges as a distinct and noteworthy area of focus. The Presidency of the Republic of Turkey's Finance Office has adopted a strategic vision with the objective of establishing Turkey as a regional leader and a global player in the field of financial technologies. In alignment with this vision, the office provides support for the advancement of the fintech ecosystem through the formulation of a range of policies and regulations. In this context, the "Turkey Fintech General Outlook" shown in Figure 4 was first published in November 2021 in coordination with the Presidency Finance Office. The objective was to provide an overview of the current status of the sector and to illustrate its development in comparison to previous years. The objective of this study is to create an up-to-date and categorized data set, providing up-to-date data on Turkey's fintech landscape for entrepreneurs and investors at home and abroad. Concurrently, the National Fintech Strategy, Turkey Fintech Guide, and Fintech Dictionary were developed in parallel with the general view of Turkey's fintech.

Figure 4. Fintech snapshot for Türkiye (Cumhurbaşkanlığı Finans Ofisi, 2023)

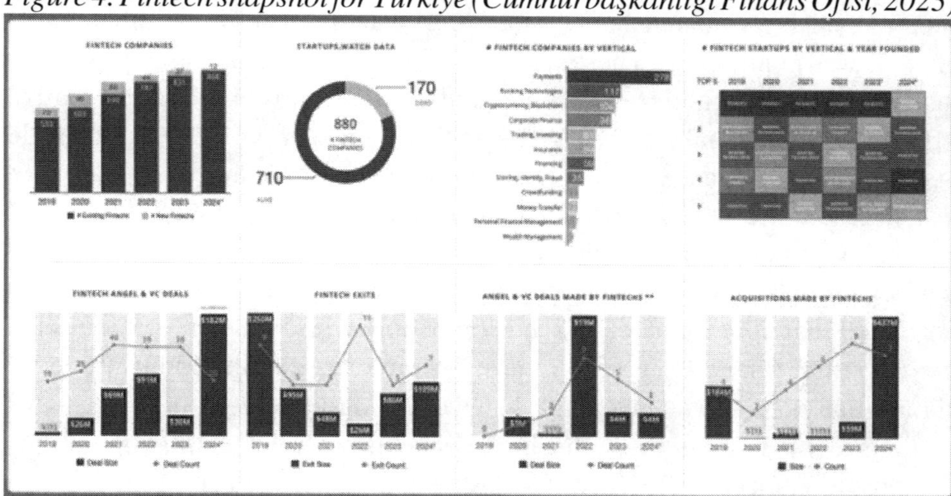

In particular, the application of technologies such as blockchain, smart contracts, artificial intelligence (AI), and mobile banking applications has the effect of enhancing the efficacy of Islamic financial products and facilitating the development of novel service models. Concurrently, these technologies enhance operational efficiency and facilitate the automation of financial processes, customer experience, risk management, and decision support systems. Blockchain technology provides a platform that aligns with the risk-sharing and transparency principles of Islamic finance, thereby ensuring the enhanced security and traceability of Islamic financial instruments, particularly Sukuk. The transparency and decentralization afforded by blockchain technology facilitate the provision of Islamic financial services in a more reliable manner. Furthermore, Islamic financial transactions can be automated through the use of smart contracts, which guarantee that transactions are executed in a manner that is independent of human error and in accordance with Sharia law. To illustrate, a Murabaha agreement may be programmed with a smart contract, thereby enabling the full automation of product delivery and payment terms. Artificial intelligence and big data are employed in domains such as customer profiling and risk analysis. Systems supported by artificial intelligence analyze customer behavior and provide personalized financial solutions. For instance, alternative data sources have been developed for use in credit assessments of individuals lacking bank accounts (Samar & Şimşek, 2020).

The integration of innovative technologies into Islamic fintech solutions is perceived as a viable means of extending financial services to unbanked or underserved populations. These solutions align with Islamic finance principles that prohibit interest and speculative activities, offering a compliant alternative to traditional financial services. This enables them to provide more accessible options for individuals who lack access to conventional financial services. For instance, Islamic peer-to-peer lending (P2P financing), crowdfunding, and digital investment platforms provide viable financing and investment opportunities for individuals lacking access to conventional banking services. It is estimated that approximately 1.7 billion people worldwide are not included in the formal financial system. Figure 5 shows the distribution of the population that does not receive bank services. A significant proportion of this group is located in regions with a high concentration of Muslim populations (Demirguc-Kunt et al., 2018). Islamic fintech platforms have emerged in both Islamic and non-Islamic countries, with the objective of including millions of young Muslims in the financial system and expanding financial services to the unbanked. The importance of digital financial solutions has increased with the advent of the coronavirus pandemic, and the role of Islamic fintech platforms has become even more evident in this process (Lacasse et al., 2020).

Islamic fintech solutions have the potential to facilitate economic development in these communities by integrating them into the financial system. Islamic fintech provides digital financial services such as digital wallets and mobile banking with advanced technologies such as blockchain and artificial intelligence, thereby enabling individuals lacking access to traditional financial services to become included in the financial system. Islamic fintech initiatives have facilitated access to digital payment systems for millions of individuals who lack conventional banking services. Such initiatives facilitate not only the provision of payment services but also the delivery of ancillary services such as microfinance and microcredit. In this manner, individuals are able to gain access to financial resources, which in turn stimulates economic growth.

Figure 5. Distribution of the unbanked population (Demirguc-Kunt et al., 2018)

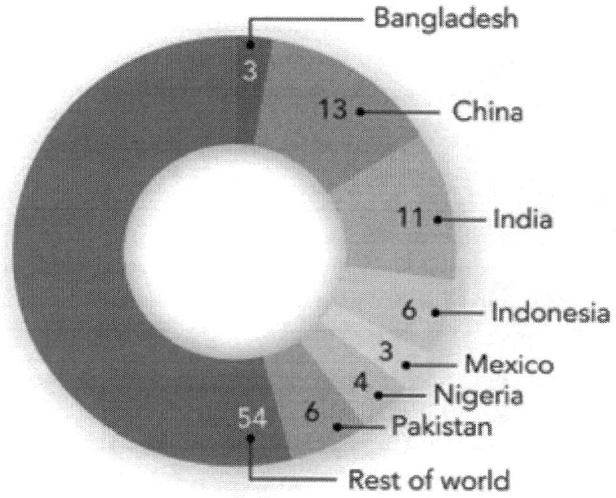

Islamic fintech plays a pivotal role in facilitating the integration of not only individuals but also small and medium-sized enterprises (SMEs) into the financial system. These businesses, which exist outside of traditional banking systems, can enhance their working capital and facilitate their growth by accessing financing through Islamic fintech solutions. Crowdfunding platforms represent an effective tool for providing entrepreneurs with interest-free financing, as part of an Islamic fintech initiative. These platforms provide entrepreneurs with access to a diverse investor base, enabling them to secure financing for their projects. At the same time, investors are able to assume a portion of the risk and make investments in alignment with ethical financing principles.

The structural deficiencies of the traditional financial system, coupled with the economic and social challenges currently facing the world, have resulted in a significant shift in consumer needs and expectations for the financial ecosystem (Emeç, 2020). In this context, Islamic fintech has facilitated the integration of new generation technologies in accordance with Islamic finance principles. This is achieved through structures that encourage innovation, thereby enhancing the inclusivity, accessibility and sustainability of financial services. By internalizing innovative technologies, they facilitate the evolution of Islamic finance in a manner that not only addresses current challenges but also shapes future global financial dynamics. This technological transformation facilitates the development of more robust and ethical financial solutions for both individuals and institutions, while simultaneously reinforcing and expanding the role of Islamic finance within the global financial system (Aktürk, 2021).

5. CONCLUSION

The advent of digital technology has opened new avenues for the advancement of Islamic finance, which is founded upon principles that encompass interest-free banking, risk distribution, transparency, and ethical finance. Moreover, the increasing integration of digital financial technologies within the global financial order is poised to play an instrumental role in this evolution. The growth of the global population, the profound transformation of the financial sector, the relentless advancement of new technologies, and the evolving customer behaviors and needs have collectively resulted in a surge in demand for Islamic finance instruments and services. The digitalization of financial services has resulted in significant advancements in terms of speed, accessibility, and security. Islamic fintech has emerged as a prominent sector within this landscape, capitalizing on the opportunities presented by the digital transformation. Technologies such as blockchain, artificial intelligence, big data, smart contracts, and mobility have enhanced the efficacy and transparency of Islamic finance, thereby facilitating the development of novel business models. Solutions based on Islamic finance principles have become attractive not only to Muslim communities, but also to a diverse customer base seeking ethical and interest-free financial solutions. This has facilitated the global expansion of Islamic fintech, offering an alternative to conventional financial solutions.

Islamic fintech has the potential to become a significant contributor to the digitalized global economy. One of the most significant advantages of Islamic fintech is its potential to enhance financial inclusion. In light of the approximately two billion individuals who are currently excluded from the financial system, a significant proportion of this population resides in Muslim-majority countries and those classified as underdeveloped. In this context, Islamic fintech has the potential to integrate these segments into the economic system by offering alternative solutions for individuals who lack access to traditional banking services. Innovative solutions, including digital wallets, peer-to-peer lending, and crowdfunding, provide viable financial services for individuals lacking access to traditional banking. These services facilitate access to finance for both individuals and small and medium-sized enterprises by offering interest-free and ethical solutions in accordance with Islamic finance principles.

Blockchain technology occupies a pivotal position within the Islamic fintech ecosystem. This technology provides a platform that is in compliance with the principles of transparency, trust, and risk sharing in Islamic finance, thereby enhancing the reliability and traceability of Islamic financial instruments, particularly sukuk. The decentralization afforded by blockchain technology facilitates the provision of Islamic financial services in a more secure manner. Transactions can be automated and executed in accordance with Islamic finance principles, free from the potential for human error, through the use of smart contracts. To illustrate, a murabaha agreement can be programmed with smart contracts, thereby completely automating product delivery and payment terms. This will facilitate the streamlining of Islamic financial processes and enhance transparency.

Islamic fintech presents a promising avenue for SMEs, offering a viable alternative to conventional financing methods. These solutions permit businesses that are not within the purview of the banking system to gain access to the capital they require without the imposition of interest and in accordance with ethical principles. Crowdfunding platforms provide support to SMEs through their extensive networks of investors, facilitating the financing of entrepreneurial projects. Furthermore, these investment vehicles provide investors with the opportunity to share risk and make responsible investments. In this context, Islamic fintech contributes to economic growth by facilitating the integration of SMEs into the financial system. Nevertheless, the Islamic fintech ecosystem is confronted with a number of challenges inherent to its developmental trajectory. The digitalization of Islamic financial products and their integration with fintech solutions necessitates meticulous monitoring of these products to ensure compliance with Islamic finance principles. It is of great importance for Islamic fintech to be more widely accepted and to grow on a global scale that there be a strengthening of cooperation with regulatory authorities and the establishment of industry standards. It is evident that regulations which are aligned with technological advancements and facilitate innovation can play a pivotal role in fostering the global growth of Islamic fintech.

In order to facilitate the development of the Islamic fintech ecosystem in Turkey, it is essential to enhance regulatory frameworks, reinforce regulations through legal backing, enhance public awareness, and bolster technological innovation. Despite Turkey's considerable potential in the domain of Islamic finance and fintech, the constraints imposed by legal and regulatory deficiencies impede the sector's growth. The advancement of Islamic fintech in Turkey hinges on the formulation of strategies that enhance its competitiveness at both the domestic and international levels. Turkey's increased investment in this field and support for the Islamic fintech ecosystem will contribute to the country's stronger position in the global fintech market, thereby enhancing its competitive advantage.

Consequently, Islamic fintech plays a pivotal role in the digital transformation of financial services, facilitating the global expansion of Islamic finance. As a consequence of the advent of innovative technologies, Islamic financial services are becoming more transparent, accessible, and reliable. Nevertheless, for the comprehensive potential of Islamic fintech to be actualized, it is imperative that regulatory infrastructures be enhanced and Sharia compliance procedures be effectively managed. In this process, the ethical and interest-free financial solutions offered by Islamic fintech provide an attractive alternative not only for Muslim communities but also for all those seeking ethical financial services within the global financial order. It seems inevitable that Islamic fintech will continue to play an important role in the digitalization of the financial system and the development of sustainable financial solutions in the future.

REFERENCES

Abojeib, M., & Habib, F. (2019). Blockchain for Islamic Social Responsibility Institutions. In *Advances in finance, accounting, and economics book series* (pp. 221–240). DOI: 10.4018/978-1-5225-7805-5.ch010

Aktürk, B. (2021). *İslami finansta finansal teknoloji (FinTek) ve FinTek'in katılım bankaları uygulamaları*. https://openaccess.marmara.edu.tr/items/e6649a77-5d1f-42fa-834a-d34c7af2d989

Aktürk, B. (2024). İslami Fintek:Yenilikçi Finans Çözümleri in *İKAM ARAŞTIRMA RAPORLARI* Vol. 32, İLKE Yayın. https://ikam.org.tr/images/fintek_raporu/ikam_arastirma_raporu_32.pdf

Alam, N., Gupta, L., & Zameni, A. (2019). Fintech and Islamic Finance. In *Springer eBooks*. DOI: 10.1007/978-3-030-24666-2

Alam, N., Gupta, L., & Zameni, A. (2021). *Fintech ve İslami Finans; Dijitalleşme, Kalkınma ve Yenilikçi Yıkım*. Albaraka Yayınları.

Cengiz, S., & Özkan, T. (2023). The Place of FinTech Applications in Islamic Finance: A Conceptual Evaluation. *Journal of Ilahiyat Researches*, *60*(1), 1–14.

Cumhurbaşkanlığı Finans Ofisi. (2023). *Fintek sözlüğü*. cbfo.gov.tr/

Demirdöğen, Y. (2023). Blockchain ve İslami Finans: Türkiye'de Uygulamalar ve Gelecek Perspektifleri. *Türkiye İslami Fintech Araştırmaları Dergisi*, 30-48.

Demirguc-Kunt, A., Klapper, L., Singer, D., Ansar, S., & Hess, J. (2018). The Global Findex Database 2017: Measuring Financial Inclusion and the Fintech Revolution. In *Washington, DC: World Bank eBooks*. DOI: 10.1596/978-1-4648-1259-0

DinarStandart. (2018). "Islamic Fintech Report 2018: Current Lanscape and Path Forward." Dubai Islamic Economy Development Centre 1-38.

El-Gamal, M. A. (2006). Islamic Finance: Law, Economics and Practice. *Cambridge University Press*, 221. DOI: 10.4197/islec.21-2.5

Emeç, Ö. (2020). *Yeni Dünya ve Yeni Finans: Ortaklık Temelli Finansman ve Katılım Bankaları*. Albaraka Yayınları.

Faster Capital. (2024). *İslami Fintech: Dijital İslami Bankacılık Hizmetlerinde Yenilikler*. https://fastercapital.com/: https://fastercapital.com/content/Islamic-Fintech--Innovations-in-Digital-Islamic-Banking Services.html#:~:text=The%20utilization%20of%20blockchain%20technology,issuance%2C%20and%20Takaful%20

Finterra. (2024, January 13). Islamic FinTech – An Evolution, or a Revolution! - The Finterra Publication - Medium. *Medium*. https://medium.com/finterra/islamic-fintech-an-evolution-or-a-revolution-bf9a80b27e36

Gateway, S. (2019). *Global Islamic Fintech Report 2019 | Salaam Gateway - Global Islamic Economy Gateway*. Salaam Gateway - Global Islamic Economy Gateway. https://salaamgateway.com/specialcoverage/islamic-fintech-2019

Gateway, S. (2024). *Top 30 Digital Islamic Economy Startups 2024*. Salaam Gateway.

Gateway, S. (2024). *Global Islamic Fintech Report 2023/24 | Salaam Gateway - Global Islamic Economy Gateway*. Salaam Gateway - Global Islamic Economy Gateway. https://salaamgateway.com/specialcoverage/islamic-fintech-2023

Grand View Research. (2023). Cryptocurrency Market Size, Share & Growth Report, 2030. Retrieved May 14, 2024, from https://www.grandviewresearch.com/industry-analysis/cryptocurrency-market-report

Greggwirth. (2022). *The ESG potential of Islamic finance - Thomson Reuters Institute*. Thomson Reuters Institute. https://www.thomsonreuters.com/en-us/posts/news-and-media/islamic-finance-esg/

Hacak, H., & Gürbüz, Y. (2019). İslami Finansta Sigorta ve Katılım Sigortası (Tekâfül)", Yaşayan ve gelişen katılım bankacılığı. *TKBB (Türkiye Katılım Bankaları Birliği) Yayınları*, 300-317.

Hasan, M. B., Rashid, M. M., Shafiullah, M., & Sarker, T. (2022). How resilient are Islamic financial markets during the COVID-19 pandemic? *Pacific-Basin Finance Journal*, *74*, 101817. DOI: 10.1016/j.pacfin.2022.101817

Hud Saleh Huddin, M., Lee, M. ve Mansor, M. S. (2022). Islamic fintech nascent and on the rise. IADI Fintech Brief, 1-18.

IFN. (2024). IFNFintech on the pulse of islamic fintech, ifnfintech.com/landscape/ https://ifnfintech.com/landscape/

Karim, R. A., & Archer, S. (2013). *Islamic Finance: The New Regulatory Challenge.* Wiley Finance. https://www.wiley.com/en-ae/Islamic+Finance%3A+The+New+Regulatory+Challenge%2C+2nd+Edition-p-9781118628973

Kılıç, G., & Türkan, Y. (2023). The Emergence of Islamic Fintech and Its Applications. *International Journal of Islamic Economics and Finance Studies*, *9*(2), 212–236. DOI: 10.54427/ijisef.1328087

Lacasse, R.-M., Lambert, B., & Khan, N. (2018). Islamic Banking - Towards a Blockchain Monitoring Process. *5th International Conference on Entrepreneurial Finance* (s. 33-46). Morocco: Journal of Business and Economics.

Noronha, M. (2020, April 27). *Islamic fintech: Reaching the next generation of Muslims.* Economist İmpact: https://impact.economist.com/perspectives/financial-services/islamic-fintech-reaching-next-generation-muslims

Orhan, Z. (2023, Kasım 7). *Küresel İslami Finans Raporundan Son Durum Notları, 2023.* https://islamiktisadi.net: https://islamiktisadi.net/2023/11/07/kuresel-islami-finans-raporundan-son-durum-notlari-2023/

Oseni, U., & Ali, S. (2019). *Fintech in Islamic Finance: Theory and Practice.* Routledge. DOI: 10.4324/9781351025584

Pietro Biancone, P., Secinaro, S., & Kamal, M. (2019). Crowdfunding and Fintech: Business model sharia compliant. *European Journal of Islamic Finance*, *12*, 1–10. DOI: 10.13135/2421-2172/3260

Rabbani, M., Khan, S., & Thalassinos, E. (2020). FinTech, Blockchain and Islamic Finance: An Extensive Literature Review. *International Journal of Economics and Business Administration*, 65-86.

Samar, M., & Şimşek, M. (2020). İslami Finans Açısından Blokzincir Teknolojisi. *Necmettin Erbakan Üniversitesi Yayınları*, 81-104.

Selcuk, M., & Kaya, S. (2021). A Critical Analysis of Cryptocurrencies from an Islamic Jurisprudence Perspective. *Turkish Journal Of Islamic Economics*, *8*(1), 137–152. DOI: 10.26414/A130

Statista. (2024). *Revenue of fintech industry worldwide 2017-2028.* https://www.statista.com/statistics/1384016/estimated-revenue-of-global-fintech/

Turan, M. (2021). *Türkiye'de İslami Fintechlerin Güncel Durumu.* islamiktisadi.net: https://islamiktisadi.net/2021/02/23/turkiyede-islami-fintechlerin-guncel-durumu/

Chapter 7
The Impact of Information and Communication Technologies (ICT) on Islamic Economy and Finance

Funda Hatice Sezgin
https://orcid.org/0000-0002-2693-9601
Istanbul University-Cerrahpasa, Turkey

ABSTRACT

Digitalization has brought a significant transformation to the Islamic finance sector, enabling it to reach a wider audience. Traditional Islamic finance models were primarily based on physical branches, face-to-face customer interactions, and paper-based transactions. However, with digitalization, banking, investment, and trade processes have accelerated significantly. In particular, internet banking, mobile applications, and blockchain-based solutions have made Sharia-compliant financial transactions more accessible and secure. This transformation has increased interest in Islamic finance while also sparking new discussions on how traditional methods can be sustained in the digital environment. The aim of this study is to examine the role of ICT in Islamic finance from different perspectives and to make future-oriented

DOI: 10.4018/979-8-3693-8079-6.ch007

inferences.

INTRODUCTION

Islamic finance has been steadily gaining a stronger position in the global financial system, relying on fundamental principles such as interest-free banking, risk-sharing, and ethical investments. Information and Communication Technologies (ICT) play a critical role in making this system more efficient, transparent, and accessible. In particular, digital banking, blockchain-based smart contracts, and AI-powered financial analysis tools enable Islamic financial institutions to offer their customers faster, more secure, and Sharia-compliant services (Nabi et al., 2017). Additionally, fintech solutions allow investors and customers to assess interest-free financial products more easily and manage their funds more effectively.

The impact of ICT on Islamic finance is not limited to digitalizing services; it also plays a crucial role in enhancing financial inclusion. Mobile banking and digital wallets facilitate the integration of individuals in regions with limited access to financial services into the Islamic finance system. Furthermore, big data analytics and artificial intelligence provide more solid evaluation processes for making Sharia-compliant investment decisions (Biancone et al., 2020). ICT-based solutions enhance the competitiveness of Islamic finance in global markets by fostering the development of more innovative financial instruments that align with the principles of transparency and trust.

One of the most significant effects of digitalization on Islamic finance is the improvement of customer experience and the inclusivity of financial services. Thanks to mobile banking and internet-based financial platforms, Muslim individuals living in different parts of the world can access Sharia-compliant financial services more easily. For instance, interest-free credit systems and digital investment instruments based on profit-and-loss sharing offer new financing opportunities for both individual and corporate customers (Hendratmi et al., 2020). This development enables individuals without access to traditional financial institutions to be integrated into the financial system.

Technological advancements have also increased the transparency and reliability of Islamic finance. Distributed ledger technologies such as blockchain ensure that financial transactions in compliance with Islamic law are traceable and immutable. As a result, the collection and distribution of zakat, sadaqah, and waqf donations through digital platforms have become more secure. Moreover, blockchain-based smart contracts facilitate the automatic execution of Sharia-compliant financial agreements, reducing dependence on intermediaries. This not only lowers costs but also strengthens trust in Islamic financial institutions (Karim et al., 2022).

However, the digitalization process has also introduced certain challenges for Islamic finance. In particular, there are ongoing debates regarding the Sharia compliance of cryptocurrencies and digital assets. Some scholars argue that cryptocurrencies are not compliant with Islamic law due to their speculative and uncertain nature, while others suggest that decentralized finance may align with the interest-free finance model. Additionally, the lack of well-established regulatory frameworks for digital financial transactions creates uncertainties, especially for Islamic financial institutions operating in different countries.

In conclusion, while digitalization presents significant opportunities for the Islamic finance sector, it also brings certain ethical and regulatory challenges. The ability of digital platforms to make financial services faster, more secure, and more accessible has contributed to the global expansion of Islamic finance. However, further research is needed to determine the compliance of digital finance with Islamic law, and regulatory frameworks must be developed accordingly. In the future, it will be essential to explore how emerging technologies such as fintech and artificial intelligence can be more effectively utilized in Islamic finance.

This study will examine the significance of ICT developments in Islamic finance, the outcomes they produce, and how future perspectives will be shaped in detail.

THE IMPACT OF INFORMATION AND COMMUNICATION TECHNOLOGIES ON ISLAMIC ECONOMY AND FINANCE

FinTech and Islamic Finance: Opportunities and Challenges

FinTech has introduced innovative technologies that are revolutionizing the financial sector and creating new opportunities for Islamic finance. Traditional Islamic finance is based on principles such as interest-free banking, risk-sharing, and ethical finance. FinTech solutions support these principles by enabling more transparent, accessible, and efficient financial services. In particular, mobile banking, blockchain, artificial intelligence (AI), and digital payment systems facilitate the wider adoption of Sharia-compliant financial products. This has increased interest in Islamic finance among both individual and institutional investors (Baber, 2020).

One of the biggest opportunities FinTech offers for Islamic finance is the enhancement of financial inclusion. Individuals with limited access to traditional banking systems can benefit from Sharia-compliant financial services through digital banking and mobile wallet applications. In regions such as Africa, Southeast Asia, and the Middle East, millions of Muslims can now access interest-free loans, investment opportunities, and insurance products through digital solutions. Additionally, digital platforms allow for more effective management of Islamic social finance mechanisms such as zakat, sadaqah, and waqf (Ansori, 2019).

Blockchain technology stands out as a significant tool that supports the principles of transparency and trust in Islamic finance. Smart contracts enable the automatic execution of Islamic finance agreements, minimizing human errors and intermediary costs. Moreover, blockchain-based solutions make investment and fund management processes in line with Islamic law more reliable (Delle Foglie et al., 2021). For example, the issuance of Sukuk (Islamic bonds) has become more transparent and traceable through blockchain, making it more attractive to global investors. This contributes to the widespread adoption of Islamic finance on a global scale.

However, the integration of FinTech with Islamic finance also presents several challenges. First, there are uncertainties regarding the Sharia compliance of digital financial instruments. For example, the extent to which cryptocurrencies and decentralized finance (DeFi) applications align with interest-free finance principles is still a subject of debate. Some scholars argue that these technologies are not compliant with Islamic law due to their speculative nature and the presence of uncertainty (gharar), while others suggest that blockchain's decentralized structure aligns with the fundamental principles of Islamic economics (De Anca, 2019).

Another significant challenge is the lack of well-developed regulatory and supervisory mechanisms. The rapid development of FinTech solutions makes it difficult for regulatory bodies to oversee these technologies in a Sharia-compliant manner. Different regulatory approaches in different countries create additional compliance costs for Islamic finance institutions operating globally. Furthermore, robust cybersecurity measures are necessary to ensure that financial transactions conducted through digital platforms are transparent and secure.

In conclusion, while FinTech offers significant opportunities for Islamic finance, it also introduces various ethical, technical, and regulatory challenges. Ensuring that FinTech solutions comply with Islamic law is critical for increasing financial inclusion and making the Islamic finance sector more competitive. However, minimizing the risks associated with digitalization requires a strong regulatory framework and close collaboration between Sharia scholars and technology experts. In the future, for the Islamic FinTech ecosystem to become more sustainable, a balanced approach to technology and ethical finance must be maintained.

The Use of Blockchain and Smart Contracts in Islamic Finance

Blockchain and smart contracts are innovative technologies that make financial transactions more transparent, secure, and efficient. In Islamic finance, these technologies can support the provision of financial services in accordance with the principles of interest-free banking, risk-sharing, and ethical finance. Since blockchain offers a decentralized and immutable ledger, it enhances the traceability and transparency of financial transactions. Meanwhile, smart contracts operate automatically based on predefined rules, increasing the reliability of Islamic finance agreements. These features align well with the core principles of Islamic finance (Biancone et al., 2022).

One of the most significant impacts of blockchain technology on Islamic finance is its ability to enhance transparency and eliminate trust issues. In traditional financial systems, transactions are controlled by centralized institutions. However, with blockchain, all transactions are recorded on a public ledger that is accessible to everyone. This aligns with the transparency principle of Islamic finance. For example, managing zakat, sadaqah, and waqf donations through a blockchain-based system ensures that funds reach their intended recipients, minimizing the risk of misuse (Abojeib and Farrukh, 2021).

Smart contracts enhance the reliability and efficiency of Islamic finance agreements by ensuring their automatic execution in a Sharia-compliant manner. Traditional Islamic finance contracts often undergo lengthy and complex verification processes. However, smart contracts enable transactions such as murabaha (installment sales), mudaraba (profit-loss sharing), and sukuk (Islamic bonds) to be executed automatically according to predefined rules. This provides significant time and cost advantages while reducing risks associated with human errors (Chong and Hui, 2021).

However, there are also challenges to the widespread adoption of blockchain and smart contracts in Islamic finance. First, there are ongoing debates about whether these technologies fully comply with Sharia principles. Many blockchain-based financial assets are speculative in nature, and Islamic law prohibits transactions that involve gharar (excessive uncertainty) and maysir (gambling). Additionally, the potential connection of cryptocurrencies to interest-bearing transactions has led some scholars to approach blockchain-based financial products with caution.

Another significant challenge is the lack of regulatory frameworks. For blockchain and smart contracts to gain widespread adoption in Islamic finance, both local and international regulatory bodies must establish appropriate legal frameworks. To ensure compliance with Sharia principles, greater collaboration between Islamic finance experts and technology developers is needed in the certification and supervision of blockchain-based financial products. Additionally, technical concerns such as cybersecurity risks and data protection must also be considered (Goud et al., 2021).

In conclusion, blockchain and smart contracts are powerful tools that can enhance the security, transparency, and accessibility of Islamic finance. These technologies have great potential, especially in automating and ensuring the compliance of Islamic financial products. However, for full integration into Islamic finance, regulatory mechanisms must be developed, and financial literacy must be improved. In the future, increased investment by Islamic finance institutions in blockchain and smart contracts will strengthen the sector's global competitiveness.

Big Data and AI-Powered Islamic Financial Analytics

Big data and artificial intelligence (AI) are revolutionizing the financial sector while also offering significant opportunities for Islamic finance. The core principles of Islamic finance—interest-free transactions, risk-sharing, and ethical investment—can be managed more effectively with big data and AI technologies. Big data analytics helps Islamic finance institutions better understand customer behavior, optimize risk analysis, and make more informed investment decisions. AI, on the other hand, can enhance the efficiency of Islamic financial services by automating complex financial processes and developing predictive models.

Big data analytics helps Islamic finance institutions identify customer needs more accurately. In traditional Islamic banking systems, financial behavior analysis is often based on a limited dataset. However, big data technologies allow financial institutions to process information from diverse sources such as social media, mobile applications, and digital banking. This enables the personalization of financial services, the efficient management of zakat and sadaqah donations, and the accurate analysis of Sharia-compliant investment opportunities. Additionally, AI-powered chatbots and automated customer support systems can improve customer satisfaction (Hassan et al., 2022).

AI plays a crucial role in risk management and fraud detection in Islamic finance. Since risk-sharing is fundamental in Islamic finance, AI-powered algorithms can help investors build lower-risk and Sharia-compliant investment portfolios. Moreover, machine learning models can detect fraudulent activities, enhance security, and protect customer data. The integration of blockchain and AI can facilitate the auditing and automatic execution of smart contracts, making Islamic finance operations more reliable (Hidajat, 2020).

Big data and AI are also transforming Sharia-compliant investment and financial analytics. One of the key principles of Islamic finance is ethical investment (halal investment), which big data analytics can identify more effectively. AI algorithms can analyze a company's business activities, revenue sources, and operational models to determine its compliance with Islamic finance regulations (Mohamed, 2021). For example, AI-driven analysis can automatically detect whether an investment fund operates in industries prohibited by Islamic law, such as alcohol, tobacco, or interest-based financial services. This allows investors to build Sharia-compliant investment portfolios.

However, the use of big data and AI in Islamic finance also presents challenges. The lack of clear regulations regarding these technologies creates uncertainty for financial institutions. Moreover, AI models must be developed within ethical frameworks. For example, AI-powered credit assessment processes must ensure fairness and avoid discrimination. Additionally, the protection of personal data in big data analytics is a crucial concern. Islamic finance institutions must adhere to strict data protection policies and regulatory frameworks (Rabbani et al., 2022b).

Big data and AI have the potential to enhance the transparency, security, and efficiency of Islamic finance. AI-powered financial analytics support more informed investment decisions, while big data analytics helps better understand customer behavior and mitigate financial risks. However, these technologies must be thoroughly evaluated for Sharia compliance and supported by regulatory frameworks. In the future, Islamic finance institutions must invest more in big data and AI to develop digital solutions aligned with ethical finance principles.

Mobile Banking and Digital Payment Systems

Mobile banking and digital payment systems have revolutionized the financial sector by making traditional banking services faster, more accessible, and user-friendly. From the perspective of Islamic finance, these technologies facilitate the widespread adoption of Sharia-compliant banking services and enhance financial inclusion. With the help of mobile applications, customers can easily access interest-free banking services and manage Islamic financial mechanisms such as zakat and sadaqah in a digital environment. This transformation has contributed to the global expansion of Islamic finance (Rizal and Rofiqo, 2020).

Mobile banking enables the digitization of traditional banking transactions, allowing customers to manage their accounts, transfer money, and pay bills without the need to visit a physical branch. For Islamic banks, this digital transformation has created an opportunity to make interest-free banking products more accessible. Financial transactions such as mudaraba (profit-loss sharing), murabaha (installment sales), and sukuk (Islamic bonds) can now be carried out more quickly and transparently through mobile platforms. Additionally, mobile banking applications can provide informative content to customers regarding Sharia-compliant investment opportunities (Qudah et al., 2023).

Digital payment systems have also brought significant changes to the Islamic finance sector. Compared to traditional payment methods, digital payments offer greater security, speed, and lower transaction costs. Mobile wallets, QR code-based payments, and blockchain-based transfers can be designed in compliance with interest-free finanee principles and integrated into the Islamic finance ecosystem. Particularly in the e-commerce sector, the widespread use of digital payment systems provides a significant advantage in promoting Sharia-compliant trade. For example, online platforms selling halal-certified products can offer interest-free payment options, ensuring compliance with Islamic financial regulations (Siska, 2022).

Another key benefit of mobile banking and digital payment systems is their role in enhancing financial inclusion. Individuals who lack access to traditional banking services can now easily access financial services through mobile banking. This is particularly beneficial for low-income individuals and Muslim communities living in rural areas, as mobile banking improves the accessibility of Sharia-compliant financial services and contributes to economic development. Additionally, these systems make the collection and distribution of zakat and charitable donations more transparent and efficient, enabling charitable organizations to reach a wider audience through digital platforms (Rosyadah et al., 2020).

However, integrating mobile banking and digital payment systems with Islamic finance presents certain challenges. First and foremost, clear regulatory frameworks need to be established to ensure the Sharia compliance of these technologies. The algorithms used in digital payment systems must be designed in accordance with interest-free financial principles. Additionally, cybersecurity risks and data protection remain major concerns in digital finance. Islamic financial institutions must implement strong measures to protect customer data and ensure security within digital financial systems.

In conclusion, mobile banking and digital payment systems serve as powerful tools for expanding the reach of Islamic finance. These technologies enhance the accessibility of financial services while promoting interest-free banking systems that align with Islamic principles. However, strengthening ethical and regulatory frameworks, improving cybersecurity measures, and developing Sharia-compliant financial products are crucial steps in this transformation process. In the future, Islamic financial institutions must invest more in mobile and digital payment systems to enhance their competitiveness in the industry.

ISLAMIC BANKING AND ICT INTEGRATION

Sharia-Compliant Digital Banking Solutions

Sharia-compliant digital banking solutions integrate traditional Islamic banking principles with modern financial technologies, providing interest-free and ethical financial services in a digital environment. The core principles of Islamic finance include the prohibition of interest (riba), avoidance of uncertainty (gharar), and the encouragement of ethical investments. Digital banking preserves these principles while offering customers faster, more secure, and more accessible financial services. Technologies such as mobile banking, blockchain-based solutions, and smart contracts allow Islamic financial institutions to improve the efficiency of their services (Oberauer, 2018).

One of the most significant aspects of Sharia-compliant digital banking is the accessibility of interest-free financial products in digital environments. While conventional banking systems rely heavily on interest-based loans, Islamic banks use alternative financing models such as mudaraba (profit-loss sharing), murabaha (installment sales), and ijara (leasing contracts). Digital banking enhances the transparency and efficiency of these transactions, allowing customers to access interest-free banking services instantly. For example, a customer can apply for murabaha-based financing through a mobile application and track the approval process digitally (Nurhadi, 2019).

Blockchain and smart contracts play a crucial role in increasing the reliability of Sharia-compliant digital banking solutions. Smart contracts enable Islamic banking transactions to be automatically executed according to Sharia principles, minimizing errors and human intervention. For example, mudaraba-based investment agreements or zakat management systems can be made transparent and traceable using blockchain technology. These innovations ensure secure and ethical financial services for both individual and corporate customers (Truby and Otabek, 2022).

Another advantage of digital banking solutions is their role in promoting financial inclusion. Individuals who lack access to traditional Islamic banks can now benefit from Sharia-compliant financial services through digital banking. In developing countries, mobile banking applications and digital wallets offer millions of Muslims an opportunity to access banking services. Additionally, Islamic financial institutions can expand globally through digital platforms, reaching Muslim communities in different regions (Zuhroh, (2021).

Some challenges remain in implementing Sharia-compliant digital banking solutions. First, regulatory frameworks need to be clearly defined to ensure compliance with Islamic financial principles. Digital financial services offered by Islamic finance institutions require expert regulatory oversight to verify their adherence to Sharia law. Furthermore, technologies such as artificial intelligence (AI) and big data analytics used in digital banking must be managed ethically and fairly. Protecting customer data and ensuring cybersecurity are also critical factors for the success of digital Islamic banking.

In conclusion, Sharia-compliant digital banking solutions combine the opportunities presented by modern technologies with the ethical and interest-free principles of Islamic finance, leading to a significant transformation in the financial sector. Digitalization enhances the accessibility, transparency, and security of Islamic financial services, increasing financial inclusion and extending the reach of interest-free finance models. However, strengthening regulatory frameworks, improving Sharia oversight mechanisms, and enhancing digital security measures are essential for a successful transition. In the future, Islamic finance institutions must invest more in digital banking solutions to drive innovation in this field.

Interest-Free Banking Models in the Digitalization Era

The digitalization era has significantly transformed the financial sector, and interest-free banking models have also been impacted by these changes. Traditional interest-free banking models relied heavily on physical branch-based services, but with the rise of digitalization, these services have largely transitioned to mobile and internet-based platforms. This transition has made interest-free banking transactions faster, more cost-effective, and more accessible. Technologies such as mobile banking, blockchain-based smart contracts, and big data analytics are helping to enhance the efficiency of interest-free banking (Siska et al., 2021).

Interest-free banking models are adapting to digitalization by incorporating various technological advancements to better meet customer needs. Financial products such as murabaha, mudaraba, musharaka and ijara can now be efficiently managed through digital platforms. For example, customers can apply for murabaha-based financing through mobile applications and track their transactions digitally. This shift reduces operational costs for banks while improving the customer experience.

Blockchain technology plays a crucial role in the digital transformation of interest-free banking models. The transparency and security offered by blockchain allow interest-free finance contracts to be traceable and verifiable. With smart contracts, investment agreements such as mudaraba or musharaka can be automatically executed without the need for intermediaries. This aligns with the fundamental principles of ethical and trustworthy finance in Islamic banking and enhances transparency. Moreover, blockchain-based payment systems allow Islamic financial institutions to expand into global markets more efficiently (Zulkhibri, 2019).

Interest-free banking in the digitalization era enhances financial inclusion by providing banking services to individuals who previously lacked access. In developing countries, millions of Muslims now benefit from digital banking services that comply with Islamic finance principles. Mobile banking applications, digital wallets, and QR code-based payment systems allow individuals without access to physical bank branches to complete banking transactions easily. Additionally, the digital collection and distribution of zakat, sadaqah, and waqf donations improve the efficiency of Islamic social finance mechanisms (Saba et al., 2019).

Some challenges remain in the digital transformation of interest-free banking. One major concern is the lack of clear regulations governing digital financial services in compliance with Islamic principles. Furthermore, AI-powered credit evaluation processes must be designed ethically to ensure fairness and avoid bias. Protecting customer data privacy, preventing cybersecurity threats, and ensuring compliance with Islamic finance ethics are crucial to the sustainability of interest-free banking.

Digitalization has the potential to make interest-free banking models more accessible, transparent, and secure. However, strengthening regulatory frameworks, enhancing Sharia compliance monitoring, and improving digital security measures are essential for success. Islamic financial institutions must continue investing in modern technologies to expand financial inclusion and offer innovative, Sharia-compliant digital finance solutions.

Transparency and Reliability with Blockchain Technology

Blockchain technology is an innovation that can significantly contribute to the fundamental principles of Islamic finance, such as transparency, reliability, and ethical financial practices. The Islamic finance system is based on interest-free (riba-free) transactions, fair risk-sharing, and the avoidance of speculative elements (gharar and maysir). While traditional financial systems often struggle with transparency and trust issues, blockchain technology offers solutions to these challenges. The decentralized and immutable nature of blockchain enhances the reliability of Islamic financial institutions' transactions, increasing trust among investors and customers (Lahmiri and Bekiros, 2019).

Blockchain technology provides a powerful tool to enhance the transparency of services offered by Islamic financial institutions. Since transactions recorded on the blockchain cannot be altered and can be verified by all stakeholders, it becomes easier to monitor the financial flows of institutions such as Islamic banks, zakat collection organizations, and investment funds. For example, the utilization of zakat and sadaqah donations can be directly tracked by donors through blockchain-based systems. This transparency prevents misuse and strengthens the reliability of financial transactions in accordance with Islamic law (Delle Foglie et al., 2021).

Another important aspect is that blockchain-based smart contracts enhance the security of Islamic finance agreements. In traditional finance systems, executing Islamic finance contracts often requires lengthy procedures and intermediary institutions. However, smart contracts on the blockchain enable the automatic and reliable execution of Sharia-compliant financial transactions. For example, financial agreements based on mudaraba (profit-loss sharing) or musharaka (joint investment) can be securely and transparently implemented through blockchain (Chong and Hui, 2021).

Blockchain technology can also increase transparency and reliability in the sukuk (Islamic bonds) market. Sukuk is a financing instrument that allows investors to participate in asset ownership without interest. However, traditional sukuk issuance processes often face information asymmetry and uncertainty (gharar). Blockchain-based sukuk systems allow investors to monitor the entire process in real time, improving transparency in transactions. Additionally, blockchain technology enables investors to see exactly where their funds are being allocated, ensuring their investments comply with Islamic principles (Abojeib and Farrukh, 2021).

Blockchain technology presents a significant opportunity for Islamic microfinance and crowdfunding platforms as well. In traditional finance systems, fundraising processes often face high transaction costs and a lack of transparency. Blockchain-based platforms, on the other hand, can manage donation or investment processes directly between investors and entrepreneurs, ensuring reliability. In interest-free microfinance projects, blockchain technology can facilitate direct connections between lenders and borrowers, reducing transaction costs and increasing trust in the system (Chong and Ling, 2021).

However, integrating blockchain technology into Islamic finance comes with challenges. First, uncertainties remain regarding the Sharia compliance of cryptocurrencies and decentralized finance (DeFi) applications. Some scholars argue that blockchain-based financial assets may not be Sharia-compliant due to their speculative nature, while others suggest that these technologies offer a major opportunity to create a fair and interest-free financial system. Therefore, Sharia-compliant blockchain applications must be designed, and regulations should be developed accordingly.

Blockchain technology has the potential to enhance transparency and reliability in the Islamic finance sector. Through blockchain, Islamic financial institutions can provide customers with more secure, traceable, and fair financial services. The automation of smart contracts, the increased transparency of sukuk issuance, and the enhanced security of microfinance systems are among the key opportunities blockchain offers. However, the use of this technology must comply with Islamic legal principles, and appropriate regulatory frameworks must be developed. In the future, it is expected that Islamic finance institutions will further adopt blockchain technology to strengthen the principles of interest-free and ethical finance.

The Impact of Artificial Intelligence and Automation on Islamic Banking

Artificial intelligence and automation are revolutionizing the financial sector, and Islamic banking is significantly influenced by these technologies. Islamic banking is based on interest-free financial principles, risk-sharing, and ethical finance. AI and automation support these principles by making

Islamic banking more efficient, accessible, and reliable. AI-powered solutions offer great advantages to Islamic financial institutions in areas such as customer service, risk analysis, fraud detection, and investment management.

One of the biggest contributions of AI and automation is improving customer service efficiency. Traditional Islamic banking is based on direct, personalized customer relationships, whereas AI-powered chatbots and digital assistants enhance service quality by providing customers with instant and accurate information. For instance, if a customer wants to learn about murabaha or mudaraba an AI-driven system can offer personalized recommendations, speeding up the process. This increases customer satisfaction while reducing operational costs for banks (Chowdhury and Uddin, 2021)

AI also plays a significant role in risk management and fraud detection in Islamic banking. Sharia-compliant financial transactions must avoid uncertainty (gharar) and excessive risk. AI algorithms analyze big data to assess customer profiles and provide financial recommendations that minimize risks. Additionally, AI-driven security systems can detect fraudulent and illegal financial transactions in real time, enhancing the credibility and transparency of Islamic banking (Hasnan, 2019).

Another advantage of AI for Islamic financial institutions is in investment management and financial decision-making. Sharia-compliant investment funds and sukuk (Islamic bonds) must adhere to specific ethical principles. AI-powered investment analysis systems can analyze financial data and identify the most suitable and Sharia-compliant investment opportunities for investors. For example, AI algorithms that examine a company's operations can determine whether it generates interest-based revenue, helping investors make informed decisions. This supports Islamic banks in maintaining their ethical finance commitments (Anindyastri et al., 2022).

Automation significantly contributes to accelerating processes and reducing costs in Islamic banking. Traditional Islamic finance transactions often require lengthy and complex verification processes. However, AI-driven automation systems ensure that financial processes are executed faster and more accurately. For instance, the distribution of zakat and sadaqah funds can be carried out more fairly and transparently through automation systems. Additionally, AI-powered automation improves operational efficiency by reducing manual workloads in banking transactions (Altarturi et al., 2022).

The integration of AI and automation into Islamic banking presents several challenges. First, AI systems must be specifically designed to comply with Sharia-compliant financial principles. Traditional AI models are often built around interest-based financial systems, requiring the development of algorithms tailored to Islamic finance principles. Additionally, ensuring that AI-powered decision-making mechanisms are transparent is essential for maintaining trust in Islamic finance.

Moreover, the ethical and social impact of AI-driven financial systems must be considered. Islamic banking aims to promote social welfare, not just economic gains. Therefore, AI systems must adhere to ethical finance principles, ensuring fair credit evaluations and non-discriminatory financial services. Protecting customer data privacy and addressing cybersecurity threats are also critical factors in ensuring the sustainability of digital Islamic banking.

AI and automation offer significant opportunities for Islamic banking, but their implementation must be approached with caution. These technologies enhance efficiency and reliability in areas ranging from customer service to investment management, risk analysis, and process automation. However, developing Sharia-compliant AI algorithms, maintaining ethical finance principles, and strengthening regulatory oversight are crucial for success. In the future, Islamic banks are expected to adopt AI and automation technologies more extensively, contributing to a more transparent and sustainable interest-free financial system.

REGULATIONS, ETHICS AND ICT COMPLIANCE WITH SHARIA PRINCIPLES

Sharia Compliance of Digital Financial Services

The rapid expansion of digital financial services presents great opportunities for Islamic finance while also raising critical questions regarding their compliance with Sharia principles. Islamic finance is based on fundamental principles such as the prohibition of interest (riba), avoidance of uncertainty (gharar), risk-sharing, and ethical finance. Digital financial services can be adapted to align with these principles, but regulatory frameworks and oversight mechanisms must be developed to ensure full compliance with Islamic law (Al-Eitan et al., 2021).

Digital banking is one of the fastest-growing areas of integration within the Islamic finance system. While traditional Islamic banking was primarily branch-based, digital solutions such as mobile banking and online banking have made interest-free financial services more accessible. For example, murabaha, mudaraba and musharaka are Sharia-compliant financial products that can now be delivered more efficiently through digital platforms. However, to ensure their compliance with Islamic principles, these services must not involve interest and must be based on fair risk-sharing (Daryanto et al., 2020).

Blockchain and smart contracts offer great potential for Sharia-compliant digital financial services. Blockchain technology ensures that transactions are transparent, immutable, and traceable, which aligns with the core principles of Islamic finance. In particular, zakat, sadaqah, and waqf donations can be managed through blockchain-based systems, ensuring that funds reach the intended recipients without misuse. Smart contracts can automate the execution of Islamic finance agreements, minimizing human errors and intermediary costs. However, the speculative nature of blockchain-based crypto assets has sparked ongoing debates regarding their compliance with Sharia principles (Hassan et al., 2020).

Digital payment systems and mobile wallets are also driving significant progress in Islamic finance. Sharia-compliant digital payment systems must be interest-free and aligned with Islamic business models. Halal-certified e-commerce platforms can integrate Sharia-compliant payment methods, allowing consumers to shop securely under Islamic finance principles. Ensuring that digital payment systems remain transparent and free of interest is essential for the development of Sharia-compliant digital financial services (Haqqi, 2020).

Artificial intelligence and big data analytics can play a crucial role in Sharia-compliant financial decision-making. Islamic financial institutions can develop AI-driven risk analysis and investment management systems to offer customers ethical and interest-free investment opportunities. For instance, AI algorithms can analyze whether specific companies or funds comply with Islamic financial principles, guiding investors accordingly. However, AI-powered financial systems must adhere to ethical standards, ensuring fairness, non-discrimination, and the protection of customer privacy (Kuanova et al., 2021).

Nevertheless, several challenges remain in ensuring the Sharia compliance of digital financial services. One of the main concerns is the lack of clear and standardized Sharia regulations in the digital finance sector, which creates uncertainty. Each digital financial product must undergo a rigorous review to determine its compliance with Islamic law. Additionally, to ensure that digital financial services can be implemented globally, international Islamic finance organizations must establish common standards. Cybersecurity threats and risks such as digital fraud must also be carefully addressed to ensure the sustainability of digital Islamic finance.

Ensuring the Sharia compliance of digital financial services is a complex but essential process. Aligning digital financial tools with Islamic law will not only increase financial inclusion but also safeguard the ethical and interest-free principles of Islamic finance. Technologies such as blockchain, AI, and digital payment systems can enhance the efficiency of Islamic finance, but they must be regulated appropriately to comply with Sharia principles. In the future, Islamic financial institutions must invest more in digital transformation, developing ethical and reliable digital financial solutions that align with Islamic finance values (Oladapo et al., 2022).

The Impact of ICT on the Financial Sector from an Islamic Law Perspective

Information and Communication Technologies have transformed the financial sector by making financial transactions faster, more secure, and more accessible. From the perspective of Islamic law, the impact of these technologies on the financial sector must be examined within the framework of interest-free financial principles, risk-sharing, and ethical finance. Islamic finance is based on fundamental principles such as the prohibition of interest (riba), avoidance of speculation (maysir) and uncertainty (gharar). Innovative ICT solutions, including digital banking, mobile payment systems, blockchain, and artificial intelligence, can support the execution of Islamic financial services in compliance with Sharia principles (Alshater et al., 2020).

One of the most significant impacts of ICT on Islamic finance is enhancing financial accessibility. Traditional Islamic banking systems primarily operated through physical branches, whereas digital banking has allowed interest-free banking products to reach a wider audience. Mobile banking applications facilitate various services, from account management to digital payment systems, while also accelerating Sharia-compliant financial transactions. As a result, Muslim communities in different parts of the world can benefit from interest-free banking services more easily (Narayan and Phan, 2019).

Blockchain technology is an essential innovation that enhances the applicability of Islamic law in the financial sector. By ensuring that transactions are decentralized and immutable, blockchain provides transparency and reliability in financial transactions. Specifically, Islamic social finance mechanisms, such as zakat and sadaqah, can be managed via blockchain-based systems, ensuring that funds are collected and distributed transparently. Additionally, smart contracts facilitate automatic execution of mudaraba and musharaka agreements, reducing human errors and fraudulent activities (Rabbani, 2022a).

Another major advantage ICT provides for Islamic finance is the integration of digital payment systems into interest-free banking models. In conventional financial systems, payment services often involve interest-based processes, whereas Islamic financial institutions are developing new models to ensure Sharia-compliant digital payment solutions. For example, mobile wallets, QR code-based payment systems, and blockchain-supported transfer mechanisms offer transparent and secure alternatives for interest-free transactions, enabling the expansion of ethical financial services.

ICT also presents challenges from an Islamic law perspective. Regulatory frameworks defining the Sharia compliance of digital financial services have not yet been fully established. In particular, the compatibility of cryptocurrencies and decentralized finance (DeFi) applications with Islamic law remains a topic of debate. Some scholars argue that the speculative nature and uncertainty of cryptocurrencies make them non-compliant with Sharia, while others suggest that blockchain technology can be aligned with Islamic finance if developed accordingly. To resolve these uncertainties, Islamic finance experts and technology developers must work together to create appropriate regulations (Razak et al., 2021).

Another critical issue is ensuring that financial services provided through ICT comply with ethical standards. Islamic law is founded on justice and fairness, which means digital banking and AI-driven financial services must be designed to provide non-discriminatory and fair financial solutions. AI-powered credit assessments and investment analysis systems must use transparent and fair algorithms when evaluating customers' financial backgrounds. Furthermore, customer data protection and cybersecurity should be considered essential components of Islamic finance's ethical standards.

Data Security and Customer Privacy: A Sharia Perspective

Data security and customer privacy are critical concerns in modern financial systems and are directly linked to the fundamental principles of Islamic law. Sharia places great importance on individual privacy and strictly prohibits the unauthorized use, misuse, or disclosure of personal information. As digital banking, mobile payment systems, and blockchain-based financial solutions become more widespread, ensuring customer data protection and transaction security has become a significant responsibility for Islamic financial institutions. Sharia-compliant financial services must not only adhere to interest-free banking principles but also guarantee the protection of customer data (Hudaefi et al., 2023).

The importance of data privacy in Islamic law is consistent with the principles of personal rights and security mentioned in the Quran and Hadith. For example, the Quran emphasizes that people's private lives should be respected and that unauthorized access to information is forbidden. Based on these principles, Islamic financial institutions must not share customers' financial or personal data without their consent and must implement the highest security standards in data storage and processing. As digitalization increases, the risk of financial data misuse grows, making the establishment of Sharia-compliant data security policies even more crucial for Islamic financial institutions.

Data security is essential in preventing fraud, identity theft, and unauthorized access in digital financial systems. Islamic law requires financial transactions to be trustworthy, fair, and transparent. Therefore, Islamic banks offering digital financial services must protect customer data using encryption technologies, secure servers, and strong authentication mechanisms. Blockchain technology is a significant innovation in this regard. Its decentralized and immutable nature enhances trust and transparency in Islamic finance, ensuring that transactions are protected from malicious manipulation (Anindyastri et al., 2022).

To establish a Sharia-compliant financial system regarding customer privacy and data security, it is essential to develop regulatory frameworks. Many countries today enforce data security laws in digital finance, but Islamic finance requires ethical and Sharia-compliant rules. For example, similar to the General Data Protection Regulation (GDPR) in the European Union, Islamic financial institutions should establish clear policies on how they collect, store, and protect customer data. Therefore, setting international Sharia-compliant data security standards is essential for the Islamic finance sector.

With the expansion of digital finance, technologies such as artificial intelligence and big data analytics have made customer data processing more sophisticated. However, from a Sharia perspective, using customer data without consent or sharing it with third parties for commercial gain is ethically unacceptable. Islamic law prioritizes fairness and honesty in all financial transactions. Therefore, AI-driven financial systems must operate transparently and respect individual privacy rights. Customer financial histories, investment preferences, and personal information should only be used to improve financial services and should never be exploited.

Another critical concern is preventing cyberattacks in digital financial systems. Islamic financial institutions must implement advanced security protocols to prevent customer data theft and misuse. Sharia mandates the protection of individual assets and rights, making security breaches and cyber vulnerabilities in digital finance a serious issue under Islamic law. To ensure secure financial transactions, biometric authentication, multi-factor authentication, and advanced encryption techniques must be employed (Khan et al., 2022).

Data security and customer privacy are directly linked to the core ethical values of Islamic finance. As digitalization advances, Islamic financial institutions must adopt strong data security policies to provide customers with secure and transparent services. Sharia-compliant financial services must go beyond interest-free banking principles to include high standards for customer data protection. In the future, Islamic financial institutions are expected to enhance cybersecurity measures and develop internationally recognized data protection policies that align with Sharia principles. This will ensure that ethical and secure digital financial solutions continue to grow within the Islamic finance sector (Nurdin and Khaeruddin, 2020).

Regulations and the Role of Regulatory Authorities

Regulations and regulatory authorities play a critical role in ensuring that digital financial services are conducted in compliance with Sharia principles. With the acceleration of digitalization, data security, customer privacy, and ethical finance have become increasingly important in the financial sector. To establish an Islamic law-compliant financial system, financial institutions operating in this field must adhere to specific ethical and legal frameworks. Regulatory bodies work to develop both local and international regulations to create a reliable digital finance ecosystem that ensures financial services remain compliant with Sharia law (Qudah et al., 2023).

Regulations in the Islamic finance sector are based on Sharia standards to ensure compliance with interest-free finance principles. International organizations such as the Organization of Islamic Cooperation (OIC), the Islamic Financial Services Board (IFSB), and the Accounting and Auditing Organization for Islamic Financial Institutions (AAOIFI), based in Bahrain, establish oversight mechanisms to ensure that Islamic finance aligns with ethical principles. These organizations publish guidelines on the use of digital banking, blockchain technology, and AI-based financial services in accordance with Sharia principles. Special standards for integrating financial technologies (FinTech) with interest-free finance models are developed to ensure investor and customer security.

Another key responsibility of regulatory bodies is to increase transparency in digital financial services and minimize cybersecurity risks. As digital banking and payment systems continue to grow, protecting customer data and preventing financial fraud has become a major concern. Regulations define how banks and financial institutions should collect, store, and protect customer data. To maintain a Sharia-compliant financial system, customer privacy must be respected, and personal data must not be shared with third parties without consent. Therefore, Islamic financial institutions must develop policies that comply with international data protection standards, such as the General Data Protection Regulation (GDPR) of the European Union (Siska, E. (2022).

Regulatory authorities also have a significant role in overseeing the integration of blockchain technology and smart contracts into the Islamic financial system. Cryptocurrencies and decentralized finance (DeFi) applications remain controversial topics in terms of Sharia compliance. Some scholars argue that blockchain technology supports transparency and traceability, making it compatible with Islamic finance, while others contend that the speculative and uncertain nature of cryptocurrencies contradicts Sharia law. Regulators must therefore establish clear rules on how Islamic financial institutions can utilize blockchain-based financial products, ensuring that these systems align with interest-free finance principles.

Regulatory bodies are also working to ensure that Islamic banks use AI and big data analytics in compliance with ethical principles. AI-driven credit assessment systems and risk analysis tools improve the efficiency of financial transactions but also pose risks of bias and unfair decision-making. Since justice and fairness are fundamental principles of Islamic finance, these technologies must be subject to ethical oversight. To ensure that AI-based financial systems provide fair financial services and do not discriminate against individuals, Sharia auditors and technology experts must collaborate to develop appropriate regulations (Abedifar et al., 2016).

Regulations and regulatory authorities play a crucial role in ensuring Sharia compliance in digital financial services. In the digitalized financial sector, there is a strong need for a regulatory framework that enhances transparency, protects customer privacy, and ensures ethical financial services. For the Islamic finance sector to thrive, it is essential to develop and continuously update regulatory mechanisms at both national and international levels. In the future, more comprehensive regulations will be necessary for Islamic financial institutions to manage digital transformation in a sustainable and ethical manner (Rabbani, 2022a).

THE FUTURE OF ISLAMIC FINANCE AND ICT

The Evolution of the Islamic Economy in a Digitalized World

As digitalization transforms the global economy, it also significantly impacts the development and operations of the Islamic economy. Traditional Islamic economics is based on interest-free banking, risk-sharing, and ethical finance principles. The digitalization process allows these fundamental principles to be delivered to a broader audience and enhances the transparency of financial systems. Emerging technologies such as blockchain, AI, big data analytics, and digital payment systems have the potential to make the Islamic economy more efficient. However, this transformation must be managed in compliance with Islamic law.

One of the greatest benefits of digitalization for the Islamic economy is the increase in financial inclusion. Millions of Muslims who lack access to conventional financial systems can now benefit from interest-free banking services through mobile banking and digital finance platforms. Particularly in developing countries, digital financial solutions have expanded the reach of Islamic microfinance and interest-free credit systems, contributing to economic development. Mobile wallets and digital payment systems simplify daily financial transactions, fostering new business models that align with the fundamental principles of the Islamic economy.

Blockchain technology plays a key role in the evolution of the Islamic economy. By enhancing the transparency and reliability of financial transactions, blockchain strengthens the ethical principles of Islamic finance. Managing zakat, sadaqah, and waqf funds through blockchain-based systems allows donors to transparently track how their funds are utilized. Additionally, smart contracts powered by blockchain enable automatic execution of mudaraba (profit-loss sharing) and sukuk (Islamic bonds), minimizing trust issues in the Islamic economy (Oseni and Nazim, 2019).

New Technologies and the Sustainability of Islamic Finance

Emerging technologies are reshaping the financial sector, offering significant opportunities for the sustainability of Islamic finance. Islamic finance is based on interest-free banking, risk-sharing, ethical finance, and social responsibility principles. Digitalization, AI, blockchain, big data analytics, and decentralized finance (DeFi) can help make Islamic finance more inclusive, transparent, and reliable. However, these new technologies must be regulated and used in compliance with Islamic law.

Blockchain and smart contracts play a critical role in the sustainability of Islamic finance. Sharia-compliant financial transactions require strict rules and oversight mechanisms. Blockchain technology ensures that transactions occur transparently and immutably, aligning with the core principles of Islamic finance. The management of sukuk (Islamic bonds) and mudaraba (profit-sharing agreements) through blockchain can increase investor confidence and streamline financial processes. Additionally, smart contracts can automate Islamic finance agreements, reducing human errors and fraudulent activities (Legowo et al., 2021).

AI and big data analytics also play a crucial role in the sustainability of Islamic finance. AI-powered systems analyze customer financial history and offer interest-free and Sharia-compliant investment opportunities. Big data analytics help Islamic financial institutions with risk management, market analysis, and understanding customer needs, enabling them to provide more efficient services. For example, AI-supported analysis systems can evaluate whether a company complies with Islamic law, making it easier for investors to identify halal investment opportunities (Kok et al., 2022).

Decentralized finance (DeFi) and crypto assets present both opportunities and risks for the sustainability of Islamic finance. DeFi reduces reliance on traditional banking systems, enabling interest-free financial transactions to reach a broader audience. However, since many crypto assets and DeFi protocols are speculative, their compliance with Sharia law remains a topic of debate. To adhere to Islamic finance principles, crypto assets must not involve speculation or uncertainty (gharar). Therefore, blockchain-based digital financial solutions must be adapted to Islamic finance principles, necessitating new regulations and fatwas (Karim et al., 2019).

For new technologies to support the sustainability of Islamic finance, regulatory frameworks and oversight mechanisms must be strengthened. Organizations such as the Islamic Financial Services Board (IFSB) and the Accounting and Auditing Organization for Islamic Financial Institutions (AAOIFI) should establish standards guiding the digital transformation of Islamic finance. Additionally, for the ethical use of financial technologies, Islamic finance experts, regulators, and technology developers must collaborate. AI-based financial tools must be designed to protect customer privacy and prevent discrimination in financial services.

The digitalization of Islamic finance presents both great opportunities and challenges. For Islamic finance institutions to maintain sustainability, they must effectively regulate and adapt to emerging technologies. By investing in AI, blockchain, and digital finance solutions, the Islamic finance sector can ensure its continued growth while adhering to Sharia principles.

The Role of ICT in Expanding into Global Markets

Information and Communication Technologies play a critical role in the expansion and globalization of Islamic finance. Traditionally, Islamic financial services operated locally or regionally, but digital solutions provided by ICT have enabled these services to reach a global audience. Mobile banking, digital payment systems, blockchain-based financial instruments, and AI-driven analytics facilitate the accessibility of Islamic finance institutions to international customers. This digital transformation not only promotes the adoption of interest-free banking principles worldwide but also increases global investor interest in Islamic financial products (Hudaefi, 2020).

Mobile and online banking are among the most essential ICT tools that enable Islamic financial services to reach a broader audience worldwide. Individuals with limited access to traditional banking systems can now easily access Sharia-compliant financial services through mobile banking applications and digital financial platforms. This is particularly beneficial in regions such as Africa, Southeast Asia, and the Middle East, where mobile-based interest-free financial services enable better economic integration. Additionally, the acceleration of international money transfers through digital payment systems allows Islamic financial institutions to expand their global customer base.

Blockchain and smart contracts are another key technology accelerating the global expansion of Islamic finance. Blockchain technology ensures that financial transactions are transparent, secure, and traceable, aligning with the ethical financial principles required by Islamic law. The issuance of sukuk (Islamic bonds) through blockchain enhances global investor interest in Islamic financial products. Furthermore, smart contracts enable automated execution of Islamic finance agreements, ensuring security in international transactions (Hasan et al., 2020).

Big data analytics and AI-powered financial tools help Islamic financial institutions make more strategic decisions in global markets. AI-driven algorithms analyze market trends, identifying international demand for Islamic financial products. This enables Islamic banks and financial institutions to develop new Sharia-compliant investment opportunities and gain a competitive edge in global markets. Additionally, big data analytics can analyze customer behaviors, offering personalized interest-free financial solutions tailored to the financial needs of individuals in different countries (Hamadou and Suleman, 2024).

CONCLUSION

Digitalization has the potential to make the Islamic economy more efficient, transparent and accessible. However, this process must be carried out in accordance with Islamic law, ethical finance principles must be protected and security threats must be prevented. In order for the Islamic economy to successfully manage the digitalization process in the future, it will be of great importance to make FinTech solutions compliant with Sharia law, to establish strong regulations and to increase cybersecurity measures. By adopting technological innovations, Islamic financial institutions can make interest-free and ethical finance sustainable in the digital age.

Emerging technologies hold great potential for the sustainability of Islamic finance. However, these technologies must be adapted to comply with Islamic law and developed in alignment with ethical finance principles. Technologies such as blockchain, AI, and big data analytics enhance transparency and efficiency, enabling Islamic finance to reach a wider audience. However, further research and regulation are needed to determine the Sharia compliance of DeFi and crypto assets. In the future, Islamic finance institutions must integrate these technologies sustainably into the digital world while upholding interest-free and ethical finance principles.

The expansion of digital financial services also brings cybersecurity threats. Islamic FinTech providers must apply high security standards to protect customer data. Since data security and customer privacy are also critical principles in Islamic law, strict regulatory oversight should be implemented, and compliance with international data protection regulations must be ensured. AI-driven fraud detection systems and advanced encryption technologies should be used to maximize security in Islamic financial services.

Islamic FinTech can also play a more active role in environmental sustainability and social impact investments. Financial instruments such as green sukuk (environmentally friendly Islamic bonds) can be used to finance sustainable projects. To contribute to the Sustainable Development Goals (SDGs), Islamic banks and FinTech companies should prioritize green finance projects. Additionally, digital zakat and sadaqah platforms should be used to enhance the effectiveness of social finance and support sustainable development projects.

ICT has the potential to make the Islamic finance sector more transparent, efficient, and inclusive. However, these technologies must be designed and regulated in compliance with Islamic law. The development of Sharia-compliant digital financial solutions will contribute to the global expansion of Islamic finance. In the future, Islamic finance institutions will need to embrace ICT and develop ethical and interest-free digital banking and payment systems.

Emerging financial technologies such as decentralized finance (DeFi) and digital assets may play a significant role in the future development of Islamic finance. However, to ensure their Sharia compliance, clear regulatory frameworks must be established. For long-term sustainability, Islamic FinTech must foster continuous collaboration among regulators, technology developers, and academics to ensure that new financial technologies are developed in accordance with Islamic law.

In order for Islamic FinTech to gain a greater place in the global financial system in the future, the above-mentioned policy recommendations need to be implemented. Regulating Sharia-compliant digital finance solutions, encouraging technological innovations, increasing financial literacy and integrating with sustainable development goals will enable Islamic FinTech to be successful at the global level. In this process, international collaborations should be increased and the adoption of ethical and interest-free finance should be encouraged.

ICT is one of the most crucial factors enabling Islamic finance institutions to expand into global markets. Digital banking, blockchain-based solutions, big data analytics, and AI-powered financial tools have made Islamic financial services more accessible internationally. However, to fully capitalize on these opportunities, it is essential to strengthen regulatory and oversight mechanisms. By leveraging technological advancements in a Sharia-compliant and ethical manner, Islamic finance institutions can establish a stronger presence in the global market. With the wider adoption of ICT, the role of Islamic finance in the global economy is expected to grow even further in the coming years.

REFERENCES

Abedifar, P., Ebrahim, S. M., Molyneux, P., & Tarazi, A. (2016). Islamic banking and finance: Recent empirical literature and directions for future research. In *A Collection of Reviews on Savings and Wealth Accumulation* (pp. 59–91). Wiley. DOI: 10.1002/9781119158424.ch4

Abojeib, M., & Habib, F. (2021). Blockchain for Islamic social responsibility institutions. In *Research Anthology on Blockchain Technology in Business, Healthcare, Education, and Government* (pp. 1114–1128). IGI Global. DOI: 10.4018/978-1-7998-5351-0.ch061

Al-Eitan, G. N., Alkhazaleh, A. M., Alkazali, A. S., & Al-Own, B. (2021). The Internal and External Determinants of the Performance of Jordanian Islamic Banks: A Panel Data Analysis. *Asian Economic and Financial Review, 11*(8), 644–657. DOI: 10.18488/journal.aefr.2021.118.644.657

Alshater, M. M., & Othman, A. H. A. (2020). Financial Technology Developments and their Effect on Islamic Finance Education, Journal of King Abdulaziz University: Islamic Economics, King Abdulaziz University. *Islamic Economics Institute., 33*(3), 161–187.

Altarturi, B. H. M., Altarturi, H. H. M., & Othman, A. H. A. (2021). Applications of financial technology in Islamic finance: A systematic bibliometric review. In *Artificial Intelligence and Islamic Finance*. Routledge. DOI: 10.4324/9781003171638-10

Anindyastri, R., Lestari, W. D., Sholahuddin, M. (2022). The Influence of Financial Technology (Fintech) on the Financial Performance of Islamic Banking (Study on Islamic Banking listed on the Indonesia Stock Exchange Period 2016-2020). Benefit: Jurnal Manajemen dan Bisnis, 7(1), 80-92.

Ansori, M. (2019). Perkembangan dan Dampak Financial Technology (Fintech) terhadap Industri Keuangan Syariah di Jawa Tengah. *Wahana Islamika: Jurnal Studi Keislaman, 5*(1), 32–45.

Baber, H. (2020). FinTech, crowdfunding and customer retention in Islamic banks. *Vision (Basel), 24*(3), 260–268. DOI: 10.1177/0972262919869765

Beik, I. S., & Arsyianti, L. D. (2021). Digital Technology and Its Impact on Islamic Social Finance Literacy, Springer Books, in: Mohd Ma'Sum Billah (ed.), Islamic FinTech, edition 1, pp. 429-445, Springer.

Biancone, P., Uluyol, B., Petricean, D., & Chmet, F. (2020). The Bibliometric Analysis of Islamic Banking and Finance. *Journal of Islamic Accounting and Business Research, 11*(10), 2069–2086. DOI: 10.1108/JIABR-08-2020-0235

Chong, F. H. L. (2021). Enhancing trust through digital Islamic finance and blockchain technology. *Qualitative Research in Financial Markets, 13*(3), 328341. DOI: 10.1108/QRFM-05-2020-0076

Chowdhury, A. I., & Uddin, M. S. (2021). Artificial intelligence, financial risk management, and Islamic finance. In *Artificial Intelligence and Islamic Finance* (pp. 181–192). Routledge. DOI: 10.4324/9781003171638-12

Daryanto, W. M., Akbar, F., & Perdana, F. A. (2020). Financial performance analysis in the banking sector: Before and after financial technology regulation in Indonesia (case study of Buku-iv in Indonesia for period 2013-2019). *International Journal of Business. Economics and Law, 21*(2), 1–9.

De Anca, C. (2019). Fintech in Islamic Finance: From collaborative finance to community-based finance. In *Fintech in Islamic Finance* (pp. 47–61). Routledge. DOI: 10.4324/9781351025584-4

Delle Foglie, A., Panetta, I. C., Boukrami, E., & Vento, G. (2021). The impact of the Blockchain technology on the global Sukuk industry: Smart contracts and asset tokenisation. *Technology Analysis and Strategic Management, 1*(1), 1–15.

Goud, B., Uddin, T. A., & Fianto, B. A. (2021). Islamic Fintech and ESG goals: Key considerations for fulfilling maqasid principles. In *Islamic Fintech* (pp. 16–35). Routledge.

Hamadou, I., & Suleman, U. (2024). FinTech and Islamic Finance: Opportunities and Challenges, Smolo, E. and Raheem, M.M. (Ed.) The Future of Islamic Finance, Emerald Publishing Limited, Leeds, pp. 175-188.

Haqqi, A. R. A. (2020). Strengthening Islamic Finance in South-East Asia Through Innovation of Islamic FinTech in Brunei Darussalam. In *Economics, Business, and Islamic Finance in ASEAN Economics Community* (pp. 202-226). IGI Global.

Hasan, R., Hassan, M. K., & Aliyu, S. (2020). Fintech and Islamic finance: Literature review and research agenda. [IJIEF]. *International Journal of Islamic Economics and Finance, 3*(1), 75–94. DOI: 10.18196/ijief.2122

Hasnan, B. (2019). Financial inclusion and FinTech: A comparative study of countries following Islamic finance and conventional finance. *Qualitative Research in Financial Markets, 12*(1), 24–42. DOI: 10.1108/QRFM-12-2018-0131

Hassan, K., Rabbani, M. R., & Mahmood, A. M. A. (2020). Challenges for the Islamic Finance and banking in post COVID era and the role of Fintech. *Journal of Economic Cooperation and Development, 41*, 93–116.

Hassan, K., Rabbani, M. R., Rashid, M., & Trinugroho, I. (2022). Islamic Fintech, Blockchain and Crowdfunding: Current Landscape and Path Forward. In *FinTech in Islamic Financial Institutions: Scope, Challenges, and Implications in Islamic Finance* (pp. 307–340). Springer International Publishing. DOI: 10.1007/978-3-031-14941-2_15

Hendratmi, A., Ryandono, M. N. H., & Sukmaningrum, P. S. (2020). Developing Islamic crowdfunding website platform for startup companies in Indonesia. *Journal of Islamic Marketing, 11*(5), 1041–1053. DOI: 10.1108/JIMA-02-2019-0022

Hidajat, T. (2020). Financial Technology in Islamic View. *Perisai: Islamic Banking and Finance Journal, 4*(2), 102–112. DOI: 10.21070/perisai.v4i2.465

Hudaefi, F. A. (2020). How does Islamic fintech promote the SDGs? Qualitative evidence from Indonesia. *Qual. Res. Financial Markets, 12*(4), 353–366. DOI: 10.1108/QRFM-05-2019-0058

Hudaefi, F. A., Hassan, M. K., & Abduh, M. (2023). Exploring the development of Islamic fintech ecosystem in Indonesial. a text analytics. *Qualitative Research in Financial Markets, 15*(3), 514–533. DOI: 10.1108/QRFM-04-2022-0058

Kannaiah, D., Masvood, Y., & Choudary, Y. L. (2017). Growth of Islamic banking in India: Discriminant analysis approach. *Banks Bank Syst., 12*(4), 175–188. DOI: 10.21511/bbs.12(4-1).2017.06

Karim, A. K. M. Rezaul, Hasan, M. (2019). Islamic Finance and Fintech: A Critical Review. *Journal of Islamic Accounting and Business Research*, *10*, 296–311.

Karim, S., Naeem, M. A., & Abaji, E. E. (2022). Is Islamic FinTech coherent with Islamic banking? A stakeholder's perspective during COVID-19. *Heliyon*, *8*(9), 1–6. DOI: 10.1016/j.heliyon.2022.e10485 PMID: 36110236

Khan, M. S., Rabbani, M. R., Hawaldar, I. T., & Bashar, A. (2022). Determinants of behavioural intentions to use Islamic financial technology: An empirical assessment. *Risks*, *10*(6), 1–19. DOI: 10.3390/risks10060114

Kok, S. K., Akwei, C., Giorgioni, G., & Farquhar, S. (2022). On the regulation of the intersection between religion and the provision of financial services: Conversations with market actors within the global Islamic financial services sector. *Research in International Business and Finance*, *59*, 101552. DOI: 10.1016/j.ribaf.2021.101552

Kuanova, L. A., Sagiyeva, R., & Shah Shirazi, N. (2021). Islamic social finance: A literature review and future research directions. *Journal of Islamic Accounting and Business Research*, *12*(5), 707728. DOI: 10.1108/JIABR-11-2020-0356

Lahmiri, S., & Bekiros, S. (2019). Decomposing the persistence structure of Islamic and green crypto-currencies with nonlinear stepwise filtering. Chaos, Solit. *Chaos, Solitons, and Fractals*, *127*, 334–341. DOI: 10.1016/j.chaos.2019.07.012

Legowo, M. B., Subanidja, S., & Sorongan, F. A. (2021). Fintech and bank: Past, present, and future. *Jurnal Teknik Komputer AMIK BSI*, *7*(1), 94–99. DOI: 10.31294/jtk.v7i1.9726

Mohamed, H. (2021). Managing Islamic financial risks and new technological risks. In *Artificial Intelligence and Islamic Finance* (pp. 61–76). Routledge. DOI: 10.4324/9781003171638-5

Nabi, G., Islam, A., Bakar, R., & Nabi, R. (2017). Islamic microfinance as a tool of financial inclusion in Bangladesh. *J. Islam. Econ. Banking Finance*, *13*(1), 24–51. DOI: 10.12816/0051154

Narayan, P. K., & Phan, D. H. B. (2019). A survey of Islamic banking and finance literature: Issues, challenges and future directions. *Pacific-Basin Finance Journal*, *53*, 484–496. DOI: 10.1016/j.pacfin.2017.06.006

Nurdin, N., & Khaeruddin, Y. (2020). Knowledge management lifecycle in Islamic bank: The case of syariah banks in Indonesia. *International Journal of Knowledge Management Studies*, *11*(1), 59–80. DOI: 10.1504/IJKMS.2020.105073

Nurhadi, N. (2019). The importance of maqashid sharia as a theory in Islamic economic business operations. [IJIBEC]. *International Journal of Islamic Business and Economics*, *3*(2), 130–145. DOI: 10.28918/ijibec.v3i2.1635

Oberauer, N. (2018). Money in classical Islam: Legal theory and economic practice. *Islam Law Soc.*, *25*(4), 427–466. DOI: 10.1163/15685195-00254A03

Oladapo, I. A., Hamoudah, M. M., Alam, M. M., Olaopa, O. R., & Muda, R. (2022). Customers' perceptions of FinTech adaptability in the Islamic banking sector: Comparative study on Malaysia and Saudi Arabia. *Journal of Modelling in Management*, *17*(4), 1241–1261. DOI: 10.1108/JM2-10-2020-0256

Oseni, U., & Nazim Ali, S. (2019). Fintech in Islamic finance. In *Fintech in Islamic Finance* (pp. 3–14). Routledge. DOI: 10.4324/9781351025584-1

Qudah, H., Malahim, S., Airout, R. M., Alomari, M., Hamour, A. A., & Alqudah, M. (2023). Islamic Finance in the Era of Financial Technology: A Bibliometric Review of Future Trends. *International Journal of Financial Studies*, *11*(2), 1–29. DOI: 10.3390/ijfs11020076

Rabbani, M. R. (2022a). Fintech innovations, scope, challenges, and implications in Islamic Finance: A systematic analysis. *International Journal of Computing and Digital Systems*, *11*, 579–608.

Rabbani, M. R., Sarea, A., Khan, S., & Abdullah, Y. (2022b). Ethical concerns in artificial intelligence (AI): The role of RegTech and Islamic finance. In Artificial Intelligence for Sustainable Finance and Sustainable Technology: Proceedings of ICGER 2021 1. Cham: Springer International Publishing, pp. 381–390.

Razak, D. A., Zulmi, S. R., & Dawami, Q. (2021). Customers' perception on islamic crowdfunding as a possible financial solution for the pandemic COVID-19 crisis in Malaysia. *Journal of Islamic Finance*, *10*, 92–100.

Rizal, F., & Rofiqo, A. (2020). Determinants of Sharia Banking Profitability: Empirical Studies in Indonesia 2011-2020. *El Barka: Jouranl of Islamic Economic and BUsiness*, *3*(1), 137–161. DOI: 10.21154/elbarka.v3i1.2051

Rosyadah, P. C., Arifin, N. R., Muhtadi, R., & Safik, M. (2020). Factors That Affect Savings In Islamic Banking. *AL-ARBAH:Journal of Islamic Finance and Banking*, *2*(1), 33–46.

Saba, I., Kouser, R., & Chaudhry, I. S. (2019). FinTech and Islamic finance-challenges and opportunities. *Review of Economics and Development Studies*, *5*(4), 581–890. DOI: 10.26710/reads.v5i4.887

Siska, E. (2022). Exploring the Essential Factors on Digital Islamic Banking Adoption in Indonesia: A Literature Review. *Jurnal Ilmiah Ekonomi Islam*, *8*(1), 124–130. DOI: 10.29040/jiei.v8i1.4090

Siska, E., Gamal, A. A. M., Ameen, A., & Amalia, M. M. (2021). Analysis Impact of Covid-19 Outbreak on Performance of Commercial Conventional Banks: Evidence from Indonesia. *International Journal of Social and Management Studies*, *2*(6), 8–16.

Truby, J., & Ismailov, O. (2022). The role and potential of blockchain technology in Islamic finance. *European Business Law Review*, *33*(2), 175–192. DOI: 10.54648/EULR2022005

Zuhroh, I. (2021). The impact of Fintech on Islamic banking and the collaboration model: A systematic review studies in Indonesia. *Jurnal Perspektif Pembiayaan Dan Pembangunan Daerah*, *9*(4), 301–312. DOI: 10.22437/ppd.v9i4.12054

Zulkhibri, M. (2019). Fintech and the Future of Islamic Finance: Opportunities and Challenges. *Journal of Islamic Monetary Economics and Finance*, *5*, 629–652.

ADDITIONAL READING

Ahmed, H. (2011). *Product development in Islamic banks*. Edinburgh University Press. DOI: 10.1515/9780748644889

Chapra, M. U. (2000). *The future of economics: An Islamic perspective*. The Islamic Foundation.

Gai, K., Qiu, M., & Sun, X. (2018). A survey on FinTech. *Journal of Network and Computer Applications*, *103*, 262–273. DOI: 10.1016/j.jnca.2017.10.011

Morhardt, J. E. (2010). Corporate social responsibility and sustainability reporting on the internet. *Business Strategy and the Environment*, *19*(7), 436–452. DOI: 10.1002/bse.657

Takeda, A., & Ito, Y. (2021). A review of FinTech research. *International Journal of Technology Management*, *86*(1), 67–88. DOI: 10.1504/IJTM.2021.115761

Wilson, J. A. (2014). The halal phenomenon: An extension or a new paradigm? *Social Business*, *4*(3), 255–271. DOI: 10.1362/204440814X14103454934294

KEY TERMS AND DEFINITIONS

Islamic Finance: "Islamic finance or Sharia-compliant finance is banking or financing activity that complies with Sharia (Islamic law) and its practical application through the development of Islamic economics. Some of the modes of Islamic finance include mudarabah (profit-sharing and loss-bearing), wadiah (safekeeping), musharaka (joint venture), murabahah (cost-plus), and ijarah (leasing)".

Digitalization: "Digitalization refers to the digitization of various information and content in parallel with technological developments, along with the expansion of internet usage. In addition to technological transformation, digitalization encompasses data storage and transmission, having an impact across many fields."

Information and Communication Technologies: "ICT, or information and communications technology (or technologies), is the infrastructure and components that enable modern computing. Among the goals of IC technologies, tools and systems is to improve the way humans create, process and share data or information with each other."

Chapter 8
The Role of Government in Promoting Sustainable Finance:
A Pathway to Achieving the SDGs in Arab Countries

Raed Awashreh
https://orcid.org/0000-0002-2252-0299
A'Sharqiyah University, Oman

ABSTRACT

This chapter explores the challenges hindering the advancement of sustainable finance in the Arab region, including regulatory gaps, limited awareness, and restricted access to funding. Using a qualitative approach based on secondary data, the research identifies key barriers such as inconsistent ESG reporting, inadequate financial infrastructure, and perceived investment risks. To foster a more enabling financial ecosystem, the study recommends regulatory reforms, capacity-building initiatives, and awareness campaigns. Additionally, strengthening development banks, promoting public-private partnerships, and leveraging emerging technologies like blockchain can enhance transparency and financial accessibility. Future research should assess these strategies to position the Arab region as a leader in sustainable finance.

DOI: 10.4018/979-8-3693-8079-6.ch008

1. INTRODUCTION

The Arab region is faced with an intricate array of socio-economic challenges, characterized by high unemployment rates, significant social disparities, and rising environmental deterioration (United Nations, 2024). These issues are further compounded by accelerated demographic growth, incomplete diversification in most economies, and exposure to exogenous shocks, such as volatility in the global oil market. Therefore, to tackle these imminent challenges, there is a need to embark on a fundamental shift towards sustainable development strategies—one that promotes long-term economic stability, social equity, and environmental sustainability. In this regard, sustainable finance has become a fundamental tool, providing innovative solutions to close financial gaps and promote social innovation, while aligning with the SDGs (Ukoba et al., 2024).

Moreover, sustainable finance is not just one financial approach or system, but rather one with the goal of achieving interconnected development outcomes. The practice consists of a variety of financial instruments and tools designed to spur investments generating economic, social, and environmental benefits (Dirie et al., 2023). For the Arab countries, with their severe resource limitations and development gaps, the adoption of sustainable finance is crucial for addressing pressing challenges such as climate change adaptation, poverty alleviation, and job creation. However, despite its potential, the implementation of sustainable finance in the region remains uneven. This imbalance requires more government intervention to create an enabling environment and effectively allocate resources (Raimi et al., 2024).

Furthermore, obstacles to the penetration and adoption of sustainable finance in Arab countries stem from gaps in the policy environment, financial instruments, and stakeholder coordination. Governments often lack incentives to encourage private sector participation, while conventional financing mechanisms are not adequately structured to address long-term sustainability issues. Additionally, empirical evidence and documentation of government-initiated projects supporting sustainable finance are scarce. These gaps hinder the region's progress in achieving the SDGs and building resilient, inclusive economies (Ikevuje et al., 2024).

The central objective of this chapter is to explore the pivotal role of the government sector in advancing sustainable finance throughout the Arab region. Through an analysis of prevailing policies, financial products, and collaborative mechanisms, the research aims to delineate practical measures by which governments can facilitate the implementation of sustainable finance principles. The central research question concerns: How can Arab government institutions effectively foster sustainable finance to address socio-economic issues and facilitate the realization of the Sustainable Development Goals?

This research is highly relevant for several reasons. First, it provides a comprehensive breakdown of the government's role in facilitating sustainable finance, thereby filling a critical knowledge gap in current literature. Second, it focuses on specific financial instruments, such as green bonds and blended finance, which can be tailored to suit the region's unique challenges and opportunities. Third, it draws valuable insights from successful Arab case studies, offering policy innovation and implementation standards. Finally, through an analysis of multi-stakeholder collaboration between governments, the private sector, and civil society, the study presents a comprehensive framework for sustainable development in line with global calls for inclusive and participatory governance in the attainment of the SDGs.

This study aims to contribute to the literature on sustainable finance by addressing the relevant research problem and aspirations, offering practical recommendations to policymakers and stakeholders in the Arab region. By combining theoretical discussions with empirical evidence, it seeks to highlight the profound transformational power of sustainable finance in supporting inclusive and robust development.

The chapter structure is as follows: Section 2 elaborates on the literature review and the role of government in sustainable finance. Section 3 presents the theoretical framework. Section 4 explains the methodology. Section 5 examines case studies, partnership models for sustainable development, barriers to enhancing finance, and solutions to these obstacles in the form of results and discussions. Finally, Section 6 concludes with recommendations, implications, and future research directions.

2. LITERATURE REVIEW

Governments play a pivotal role in shaping the landscape of sustainable finance, particularly in regions like the Arab world, where economic, social, and environmental challenges require coordinated and strategic interventions. By leveraging their regulatory authority, policy-making capabilities, and convening power, governments can establish an enabling environment that fosters sustainable finance practices (Falcone, 2020). This involves the development of regulatory frameworks and policies alongside the demonstration of leadership and commitment to driving the sustainable finance agenda. Consequently, a robust regulatory framework becomes the cornerstone of sustainable finance (Abdel-Meguid, Dahawy & Shehata, 2021), providing the necessary structure and guidelines for stakeholders to engage in sustainable financial practices. In Arab countries, where financial systems often lack integration with global sustainability standards, governments must actively design and implement policies that incentivize sustainable investments and mitigate associated risks (United Nations, 2018).

Moreover, clear definitions and criteria for sustainable finance are essential, and governments need to align national regulations with international benchmarks, such as the United Nations Sustainable Development Goals (SDGs) (Awashreh & Hassiba, 2025), and the Principles for Responsible Investment (PRI). Standardized definitions enable financial institutions, investors, and other stakeholders to align their activities with national development priorities and global sustainability objectives (Weber, 2018). Policymakers can further incentivize green investments through fiscal and monetary tools such as tax incentives, subsidies, and preferential interest rates for green bonds or renewable energy projects. Additionally, measures like exempting social impact bonds from taxation can promote funding for education or healthcare projects, reducing barriers to entry for private-sector participation. Central banks and financial regulators also play a crucial role by incorporating sustainability into their oversight frameworks (Porretta & Benassi, 2021). This includes mandating environmental, social, and governance (ESG) reporting, stress-testing financial institutions for climate risks, and requiring banks to allocate a portion of their portfolios to sustainable projects (Polzin et al., 2019).

Furthermore, governments can enhance sustainable finance by promoting public-private partnerships (PPPs), which are particularly relevant in Arab countries with limited public resources for addressing large-scale infrastructure and social needs. A well-regulated PPP model ensures that private-sector engagement aligns with public sustainability goals. Beyond national efforts, Arab governments can benefit from cross-border collaboration to scale sustainable finance initiatives (Hodge & Greve, 2007). Joining international coalitions, such as the Coalition of Finance Ministers for Climate Action, provides technical expertise, funding opportunities, and access to best practices from other regions. While regulatory frameworks and policies create the structural foundation for sustainable finance, government leadership and commitment are equally vital for ensuring effective implementation and fostering stakeholder confidence (Awashreh, 2025). Leadership manifests in the form of vision, coordination, and the ability to mobilize diverse actors toward common sustainability goals (Unger & Thielges, 2021).

A clear and compelling vision for sustainable finance can inspire action across sectors, integrating sustainable finance into broader economic and social development strategies. For instance, a national vision prioritizing renewable energy, circular economy principles, and social equity can serve as a blueprint for financial practices that align with sustainability goals. Leadership also requires resource allocation and accountability (Awashreh et al., 2024). Governments must commit budgetary support to sustainability initiatives, such as funding green infrastructure projects or providing seed capital for social enterprises, while establishing transparent mechanisms for tracking progress and reporting outcomes to maintain credibility. Additionally, creating awareness and building capacity among stakeholders is another critical leadership role (Awashreh & Mohamed, 2024). Governments can raise awareness about the benefits of sustainable finance, conduct training programs for financial institutions, and disseminate information about available incentives, fostering a culture of sustainability within public and private organizations (Qadir et al., 2021).

Governments can also lead by example, adopting sustainable practices in their operations and investments. For instance, issuing sovereign green bonds allows governments to fund sustainability projects directly while signaling their commitment to sustainable finance. Such initiatives in Arab countries could set a precedent for private investors and financial institutions to follow (Mertzanis & Tebourbi, 2024). This leadership role extends further to fostering multi-stakeholder collaboration, where governments create platforms for dialogue and cooperation among private-sector actors, civil society organizations, and international partners. These collaborative efforts are crucial in addressing systemic challenges that require pooled resources and expertise. For example, governments could convene roundtables to discuss financing solutions for climate adaptation in water-scarce regions (Iao-Jörgensen et al., 2024).

Moreover, effective government leadership ensures that sustainable finance initiatives address pressing social and economic disparities in the region. By prioritizing projects that benefit marginalized communities, create jobs, and enhance access to essential services, governments can embed equity considerations into their strategies, amplifying developmental impact while building public trust (Elamin, 2023). The government's role in sustainable finance is thus multifaceted, encompassing the creation of regulatory frameworks, the establishment of clear policies, and the demonstration of leadership and commitment. In the Arab region, where sustainability challenges are deeply intertwined with socio-economic realities, government action is critical in mobilizing resources and driving innovation. By fostering an enabling environment and leading by example, Arab governments can position themselves as catalysts for sustainable development, bridging the gap between financial systems and the SDGs. Through strategic interventions and coordinated efforts, governments can lay the foundation for a more resilient and inclusive future (Alhammadi et al., 2024).

Furthermore, sustainable finance instruments are pivotal in addressing the pressing socio-economic and environmental challenges in the Arab region. These tools enable the mobilization of capital toward projects that advance economic resilience, social equity, and environmental sustainability. Among the various instruments available, green bonds, social impact bonds, and blended finance models have emerged as key mechanisms with the potential to reshape the financial landscape in the region. Each instrument offers unique advantages and addresses specific aspects of sustainability, although their implementation comes with challenges and opportunities that warrant careful consideration (Rahman et al., 2023).

Green bonds represent one of the most promising financial instruments for financing renewable energy projects and other environmentally beneficial initiatives. These bonds allow governments, corporations, and financial institutions to raise capital specifically for projects that mitigate environmental risks or promote sustainable practices (Alharbi & Csala, 2020). In the Arab region, where reliance on fossil fuels remains significant, green bonds can serve as a strategic tool to transition toward cleaner energy sources, such as solar and wind power. For instance, countries like the UAE and Saudi Arabia, with their abundant solar potential, could leverage green bonds to fund large-scale renewable energy projects. This not only aligns with global climate goals but also addresses regional energy security and diversification needs.

However, the successful issuance of green bonds requires robust regulatory frameworks, transparent reporting standards, and investor confidence—areas where the Arab region must continue to evolve (Alharbi & Csala, 2020).

Social impact bonds offer an innovative solution to social challenges like poverty, education, and healthcare. These outcome-based contracts allow private investors to fund programs upfront and receive returns only if predetermined outcomes are met (Carè & De Lisa, 2019). This model shifts risk from governments to private investors and encourages innovation. In the Arab region, where socio-economic disparities persist, social impact bonds could fund educational and healthcare programs for underserved communities. Governments can facilitate these arrangements by defining clear metrics for success, ensuring accountability, and attracting investors through policy incentives. However, the model requires strong data systems and public-private partnerships (Kamalov et al., 2023).

Blended finance models attract private investment into projects that provide both financial returns and social or environmental benefits. They combine public and philanthropic capital to de-risk investments in high-risk sectors like renewable energy, sustainable agriculture, and affordable housing (Clark, Reed & Sunderland, 2018). In the Arab region, blended finance can help overcome barriers to private sector engagement in sustainability initiatives (OECD, 2020; Abalkina & Zaytsev, 2021). Governments could use concessional funding to catalyze private investment in climate adaptation projects. However, success depends on aligning diverse interests, managing risks, and ensuring equitable benefits (G20 Sustainable Finance Working Group, 2023).

Despite these opportunities, implementing these instruments in the Arab region faces several challenges. A major barrier is the lack of awareness and expertise among stakeholders about sustainable finance instruments. Capacity-building initiatives, such as training programs, are essential to bridge this gap (Zamiri & Esmaeili, 2024). Additionally, the region's financial systems need integration and standardization to attract global investors. Cultural and institutional resistance to new financial instruments also poses challenges, requiring strong government leadership and strategic communication to build trust (Al-Qahtani & Albakjaji, 2023).

However, the opportunities for sustainable finance in the Arab region are significant. Growing recognition of the need for sustainable development, coupled with international support, creates a favorable environment for innovation in finance. The region's young, entrepreneurial population also presents opportunities to align sustainable finance with development goals (Abo-Khalil, 2024). In conclusion, adopting sustainable finance instruments such as green bonds, social impact bonds, and blended finance can help address environmental and social challenges while fostering economic growth. Collaboration among governments, financial institutions, and investors is key to creating a sustainable financial ecosystem that supports long-term development objectives (The World Bank Group, 2020).

3. THEORETICAL FRAMEWORK FOR SUSTAINABLE FINANCE IN THE ARAB REGION

The study uses Sustainable Finance Theory, Institutional Theory, and Stakeholder Theory in determining the way Arab governments can promote sustainable finance in the Arab region (Raza, Alavi & Asif, 2024). Sustainable finance theory requires that financial decision-making takes into account economic, social, as well as environmental considerations, with financial products like green bonds promoting long-term economic sustainability and social well-being (Fu, Lu & Pirabi, 2023). Institutional theory explains how regulatory structures shape financial practices and behaviors, while stakeholder theory argues that financial decisions need to consider a range of stakeholders, including government agencies, financial institutions, private investors, civil society organizations, and international bodies (Berthod, 2018). The present study evaluates public-private partnerships, government-led multi-stakeholder initiatives, and the role played by financial institutions in driving sustainable finance. Through the adoption of a participatory approach, Arab governments are able to craft inclusive financial ecosystems that address socio-economic disparities while establishing trust and accountability. Table 1 present the theoretical framework

Table 1. Theoretical framework components

Theory	Key Aspects	Relevance to Sustainable Finance in Arab Region
Sustainable Finance Theory	Economic, social, and environmental integration	Explains the necessity of financial instruments such as green bonds, social impact bonds, and blended finance to achieve SDGs
Institutional Theory	Policies, regulations, governance structures	Highlights the role of government frameworks, regulatory incentives, and financial oversight in enabling sustainable finance
Stakeholder Theory	Multi-stakeholder engagement, collaboration	Examines how governments, private investors, and financial institutions can work together to overcome financial barriers

Sources: developed by the authors through consultation of Raza, Alavi, and Asif (2024), Fu, Lu, and Pirabi (2023), and Berthod (2018).

4. METHODOLOGY

By adopting a descriptive approach, the methodology utilizes secondary data, including a literature review and case study analysis. Advancing sustainable finance in the Arab region requires a multi-dimensional approach. The first step in this process is a comprehensive analysis of the current regulatory landscape, identifying gaps and areas where policies need strengthening to support sustainable finance. This includes examining the existing frameworks for green bonds, social impact bonds, and blended finance models across various Arab countries. A total of 71 articles and organizational reports, including those from the United Nations and other international bodies, were reviewed.

Qualitative methods, such as policy reviews, stakeholder interviews, and case examples, were employed to assess the effectiveness of these instruments and identify challenges faced by governments, financial institutions, and the private sector (Alhejaili, 2024). Table 2 provides a classification of the references used to develop this study.

Table 2. Classification of the references

Type of Source	Count
Articles (Journal papers)	61
Organizations Reports	12
Total	73

Procedures for Methodology

1. Descriptive Approach: The research adopts a descriptive methodology, focusing on secondary data sources such as literature reviews and case study analyses.
2. Regulatory Landscape Analysis:
   - Objective: Identify gaps in regulatory frameworks for sustainable finance in Arab countries.
   - Approach: Review existing policies for green bonds, social impact bonds, and blended finance models.
   - Data Sources: 70 articles and reports, including those from the UN and international bodies.
   - Tools: Qualitative techniques like policy reviews, stakeholder interviews, and content analysis.
3. Case Study Analysis:
   - Objective: Evaluate the impact of sustainable finance initiatives.
   - Approach: Analyze case studies from the UAE (green bonds), Egypt (sustainable agriculture), and Morocco (solar energy).
   - Data Collection:
     - Qualitative data (policy documents, stakeholder interviews).
     - Quantitative data (statistical analysis of investment flows, outcomes, and impacts).
   - Analysis: Compare case studies to identify successful strategies and challenges.
     - Participatory Approach:
   - Objective: Engage stakeholders to co-create solutions.
   - Approach: Organize workshops and roundtable discussions with governments, financial institutions, and NGOs.
   - Outcome: Collaborative identification of actionable steps for enhancing sustainable finance.

- Framework Illustration:
  - o A visual framework for sustainable finance in the Arab region, highlighting key topics, instruments, collaborations, and challenges.
  - o Figure 1 illustrates this framework.
- Reporting and Analysis:
  - o The results are structured to highlight:
    - Successful initiatives.
    - Collaborative sustainable development approaches.
    - Challenges and barriers.
    - Strategies to overcome these barriers.
- Iterative Process:
  - o Findings and recommendations are refined iteratively based on emerging data and insights.

The methodology combines qualitative and quantitative data to provide a comprehensive understanding of sustainable finance in the Arab region, offering actionable solutions. It evaluates the impact of sustainable finance initiatives through case studies from the UAE, Egypt, and Morocco, analyzing them to extract insights on successful financing mechanisms. A participatory approach engages stakeholders in identifying practical solutions, fostering collaboration to enhance sustainable finance across the region. Figure 1 illustrates the sustainable finance framework for the Arab region (Dushkova & Kuhlicke, 2024).

4. RESULTS

This section presents the findings from the analysis, highlighting key aspects of sustainable finance in the Arab region. It covers successful initiatives, collaborative approaches for sustainable development, and the challenges and barriers hindering progress. Additionally, it explores strategies for overcoming these obstacles to advance sustainable finance.

4.1 Successful Sustainable Finance Initiatives

The Arab region is experiencing a significant rise in successful sustainable finance initiatives that address pressing socio-economic and environmental challenges. This growth is largely driven by collaborations between governments, the private sector, and international partners, which have enabled the development of innovative financing mechanisms and policy frameworks to support sustainability efforts (Zaidan, Al-Saidi & Hammad, 2019). These initiatives not only contribute to economic development but also help countries transition toward greener economies by reducing carbon footprints, enhancing resource efficiency, and promoting social well-being.

One of the leading examples of sustainable finance in the region is the United Arab Emirates (UAE), which has positioned itself at the forefront of green finance. The country has actively developed its green bond and sukuk market, channeling investments into large-scale sustainable projects, including energy-efficient infrastructure, renewable energy expansion, and public transportation systems such as the Dubai Green Metro Line. By aligning its financial and regulatory frameworks with international sustainability standards, the UAE has successfully attracted global investors, reinforcing its commitment to achieving its Net Zero by 2050 initiative (Khiari et al., 2024; Benzaken et al., 2024). The government's proactive approach, combined with its robust financial sector and strategic vision, has established the UAE as a model for integrating sustainable finance into national development strategies.

Similarly, Egypt has leveraged blended finance to drive sustainable agriculture through its National Project for Sustainable Agriculture. This initiative aims to enhance agricultural productivity while promoting environmental resilience by attracting private sector investment in water-efficient irrigation systems and climate-resilient crop varieties. International financial institutions, such as the World Bank and the Green Climate Fund, have played a crucial role in supporting Egypt’s efforts, providing both funding and technical expertise to improve food security and climate adaptation measures (Kühlert et al., 2024). By integrating financial innovation with sustainable development goals, Egypt demonstrates how blended finance can facilitate long-term environmental and economic benefits.

Another noteworthy case is Morocco’s Noor Ouarzazate Solar Complex, one of the largest concentrated solar power plants in the world. The project exemplifies how public-private partnerships (PPPs) can effectively mobilize resources for sustainable energy projects. By combining international loans, government funding, and private investment, Morocco has strengthened its renewable energy sector, reducing dependence on fossil fuels and enhancing energy security. This initiative not only supports the country's commitment to increasing renewable energy capacity but also contributes to regional efforts in mitigating climate change (Kühlert et al., 2024; Benbba et al., 2024). The Noor Ouarzazate project highlights the transformative impact of sustainable finance when multiple stakeholders collaborate to implement large-scale green infrastructure.

In a similar vein, Jordan's Disi Water Conveyance Project addresses the country's persistent water scarcity through an innovative build-operate-transfer (BOT) financing model. This approach allows private sector participation in the development and operation of critical infrastructure while ensuring long-term sustainability and public access to essential resources. By leveraging international support and private investment, the project has significantly improved water supply reliability, particularly in urban areas facing severe water stress (Al-Addous et al., 2023; Marin, 2009). Jordan's experience underscores the importance of financial innovation in tackling environmental challenges, particularly in regions vulnerable to climate change and resource scarcity.

These diverse case studies provide valuable lessons for other Arab countries seeking to expand their sustainable finance initiatives. Key success factors include strong government leadership, clear and consistent policy frameworks, the adoption of innovative financing mechanisms, active community engagement, and robust regional and international partnerships. Effective coordination among these elements can help scale sustainable finance, ensuring long-term socio-economic and environmental benefits for the region (Gorelick, Cara & Kavoo, 2024; Alhejaili, 2024). As the Arab world continues to navigate the complexities of sustainable development, these experiences serve as guiding models for designing and implementing resilient financial strategies that drive green growth and social prosperity.

4.2 Collaborative Approaches for Sustainable Development

Sustainable development in the Arab region requires a concerted effort among governments, the private sector, non-governmental organizations (NGOs), and international institutions to address pressing challenges such as climate change, water scarcity, and social inequality. Given the region's diverse economic and environmental landscape, multi-stakeholder collaboration is essential to mobilize resources, drive innovation, and implement sustainable solutions that are both effective and scalable (UNDP, 2013). These partnerships integrate diverse expertise: governments establish policy and regulatory frameworks, the private sector provides financial capital and technological innovation, while NGOs contribute localized knowledge, community engagement, and advocacy efforts. The synergy among these actors not only helps overcome financial and technical constraints but also mitigates investment risks and ensures the successful implementation of financially viable and socially impactful sustainability initiatives (Reynolds, 2024).

A growing number of successful initiatives in the Arab world illustrate the benefits of multi-stakeholder engagement in sustainable development. In the United Arab Emirates (UAE), the collaboration between the government, private investors, and NGOs in projects such as Masdar City serves as a model for sustainable urban development. Masdar City, designed as a low-carbon eco-city, integrates energy-efficient buildings, smart infrastructure, and renewable energy sources to reduce carbon emissions while fostering economic growth. This initiative demonstrates how cross-sector partnerships can drive large-scale, innovative urban sustainability solutions (Griffiths & Sovacool, 2020; Saradara et al., 2023).

Similarly, in Egypt, the Ministry of Environment has partnered with NGOs, private sector companies, and international organizations to implement the Integrated Solid Waste Management Program. This program focuses on waste reduction, recycling, and sustainable disposal methods, addressing both environmental and public health concerns. By leveraging private sector investments and international funding, Egypt has developed an efficient waste management system that not only reduces pollution but also creates job opportunities in the circular economy (Ibrahim & Mohamed, 2016).

Morocco's Noor Ouarzazate Solar Complex stands out as another example of successful multi-stakeholder collaboration. This large-scale renewable energy project, one of the world's largest concentrated solar power plants, was made possible through a combination of public funding, private investments, and international financial support from institutions such as the World Bank and the European Investment Bank. By integrating private sector expertise and foreign capital with strong government leadership, Morocco has significantly expanded its renewable energy capacity, reducing its dependence on fossil fuels and strengthening energy security (Benbba et al., 2024).

The effectiveness of these multi-stakeholder partnerships lies in their ability to pool financial and technical resources while fostering knowledge-sharing and innovation. Regional platforms such as the Arab Forum for Sustainable Development serve as essential venues for policymakers, researchers, and private sector leaders to exchange insights and best practices, promoting a cohesive approach to sustainability (Eweje et al., 2020). Moreover, innovative financing mechanisms, such as blended finance, have proven instrumental in attracting private investment by de-risking projects through the combination of public and private capital (Jung, 2020).

Capacity-building efforts, supported by organizations such as the Arab League's Arab Ministerial Council for Environment, further strengthen the foundation for sustainable finance in the region. These initiatives provide training, technical assistance, and policy guidance, equipping stakeholders with the necessary skills to design and implement impactful sustainability programs (Clark, Reed & Sunderland, 2018). In addition, transparency and accountability mechanisms play a crucial role in ensuring the long-term success of sustainable finance. Governments must establish clear regulatory frameworks, track financial flows, and enforce Environmental, Social, and Governance (ESG) criteria in investment decisions. These measures help ensure that financial resources are directed toward projects with tangible social and environmental benefits, reinforcing investor confidence and fostering trust among stakeholders (Boscia et al., 2019; Ezeh et al., 2024).

In short, multi-stakeholder engagement is a critical driver of sustainable finance in the Arab region. By leveraging the strengths of various actors—government leadership, private sector investment, NGO expertise, and international institutional support—countries can address complex sustainability challenges and create innovative financial solutions tailored to their specific needs. The combination of collaborative approaches, transparent governance mechanisms, and capacity-building initiatives will be essential in scaling up sustainable finance efforts. As the region continues to navigate its transition toward sustainability, these partnerships will play a fundamental role in shaping a resilient and environmentally conscious future (Alhejaili, 2024).

4.3 Challenges and Barriers to Advancing Finance

Advancing sustainable finance in Arab countries faces significant challenges, including regulatory gaps, a lack of awareness, and limited access to financial resources. These obstacles hinder the transition towards environmentally and socially responsible investment practices, which are essential for sustainable economic development. Overcoming these barriers requires a multi-faceted approach that integrates policy reform, education, and capacity building (Alhejaili, 2024). Addressing these challenges is crucial not only for fostering economic resilience but also for ensuring that Arab nations align with global sustainability goals, such as the United Nations' Sustainable Development Goals (SDGs) and the Paris Agreement on climate change.

A primary obstacle is the regulatory gap, as many countries in the region lack clear, standardized policies that support sustainable investments. This regulatory uncertainty discourages investors and financial institutions from committing to long-term sustainability projects, as they fear unstable policy environments and inconsistent implementation (Agrawal et al., 2024). Moreover, the absence of well-defined frameworks for environmental, social, and governance (ESG) reporting further complicates the assessment of project sustainability, making it difficult to attract international investors. Without harmonized regulations for ESG disclosures, businesses may face challenges in demonstrating their sustainability commitments, leading to market inefficiencies and reduced investor confidence. Additionally, policies regulating green bonds, carbon pricing, and blended finance mechanisms remain underdeveloped in many Arab countries, limiting the availability of financial instruments tailored for sustainable projects (Agrawal et al., 2024).

Beyond regulatory challenges, a significant barrier is the lack of awareness and expertise among key stakeholders, including governments, financial institutions, and businesses. Many decision-makers in both the public and private sectors remain unfamiliar with sustainable finance principles and the long-term benefits of integrating ESG factors into financial decision-making. This lack of knowledge prevents widespread adoption of sustainable finance solutions, particularly in the private sector, where short-term financial returns are often prioritized over long-term sustainability objectives (Rahman et al., 2023). Additionally, many businesses and financial institutions in the region do not yet have the necessary technical expertise to develop sustainable investment portfolios, assess ESG risks, or implement sustainability-driven business models. Without targeted educational initiatives, misconceptions about the costs and benefits of sustainable finance persist, slowing its adoption.

Limited access to finance is another critical challenge. While global capital markets have seen a surge in sustainable investments, many projects in the Arab region struggle to secure adequate funding due to perceived financial risks and underdeveloped financial infrastructure. Small and medium-sized enterprises (SMEs) focusing on sustainable innovations face difficulties accessing traditional bank financing, as financial institutions often perceive such ventures as high-risk due to their relatively untested business models (Awashreh & Zaabanut, 2025; PwC, 2024). Furthermore, early-stage businesses in sectors such as renewable energy, sustainable agriculture, and green infrastructure often encounter high upfront costs and uncertain returns, deterring investors who prefer lower-risk, short-term profit opportunities. The lack of tailored financial products, such as concessional loans, sustainability-linked bonds, and impact investment funds, further limits the availability of capital for sustainability-driven businesses.

To address these issues, Arab countries must implement comprehensive policy reforms aimed at creating clear, well-defined regulations that promote sustainable finance. Governments should establish robust ESG frameworks that mandate corporate sustainability reporting and provide incentives for businesses to transition towards sustainable practices. Regulatory agencies must also work towards standardizing green bond issuance and developing blended finance mechanisms that facilitate public-private investments in sustainability-focused sectors (Khan et al., 2024). Strengthening legal frameworks for impact investing and social enterprises can further encourage private-sector engagement in sustainability efforts.

Education and awareness campaigns are equally vital in promoting sustainable finance across different sectors. Governments, financial institutions, and academic institutions should collaborate on initiatives that enhance knowledge of ESG investing, climate-related financial risks, and sustainable business models. These initiatives should be tailored to different stakeholder groups, including policymakers, financial professionals, corporate leaders, and entrepreneurs, to foster a widespread understanding of sustainable finance principles (Rahman et al., 2023). Public awareness campaigns can also play a role in increasing consumer demand for green products and services, encouraging businesses to adopt more sustainable practices.

Moreover, capacity building is essential to equipping financial institutions with the skills needed to assess ESG risks and develop innovative financial instruments that support sustainability. Banks and investment firms should receive specialized training on structuring green bonds, sustainability-linked loans, and other sustainable investment products. Simultaneously, governments must invest in enhancing the expertise of public sector officials responsible for designing and implementing sustainability policies. Educational institutions should also integrate sustainable finance topics into business and economics curricula, ensuring that future professionals are well-versed in sustainability-driven financial strategies (Dzinamarira et al., 2024; Martínez-Peláez et al., 2024).

A crucial strategy for advancing sustainable finance in the Arab world is strengthening the role of development banks and financial institutions in funding sustainability projects. Development banks can play a key role in reducing investment risks by offering concessional financing and guarantees that attract private-sector participation. These institutions can also provide long-term funding for essential sectors such as renewable energy, sustainable agriculture, and green infrastructure, fostering economic diversification and environmental resilience. Additionally, blended finance solutions—where public funds are used to leverage private investment—can help mitigate financial risks and encourage large-scale investments in sustainability-driven projects (Taghizadeh-Hesary & Yoshino, 2020).

Public-private partnerships (PPPs) are another essential tool for mobilizing resources and expertise to advance sustainable finance. Through PPPs, governments and private entities can collaborate on large-scale sustainability projects, such as smart cities, clean energy transitions, and water conservation initiatives. These partnerships help share risks, reduce financial burdens on governments, and ensure that sustainability efforts align with national economic development plans (Vassileva, 2022).

In conclusion, addressing the challenges of sustainable finance in the Arab region requires a coordinated, multi-stakeholder approach. Regulatory reforms, enhanced awareness, and increased financial accessibility are key to unlocking the region's potential in sustainable investing. By fostering an enabling environment through policy clarity, education, capacity building, and financial innovation, Arab nations can attract sustainable investments, promote inclusive economic growth, and contribute meaningfully to global sustainability efforts. With the right mix of policies, institutional support, and strategic collaborations, the Arab region can position itself as a leader in sustainable finance, paving the way for a resilient and prosperous future (Alhejaili, 2024).

4.4 Overcoming Barriers to Sustainable Finance

Advancing sustainable finance in Arab countries faces challenges that hinder the potential of investments for long-term economic, social, and environmental progress (UN, 2024). These barriers, such as regulatory gaps, lack of awareness, and limited access to finance, must be addressed through comprehensive strategies, including policy reform, education, and capacity building (Alhejaili, 2024). One of the primary challenges is the regulatory gap. Many Arab countries lack clear, standardized policies that incentivize sustainable investments, and inconsistent frameworks for Environmental, Social, and Governance (ESG) reporting hinder investor confidence (Agrawal et al., 2024). Additionally, another significant barrier is the lack of awareness of sustainable finance among key stakeholders, including governments, financial institutions, and businesses. Knowledge gaps persist, especially in the private sector, where short-term financial gains are often prioritized over long-term sustainability (GTR, 2024). These issues compound each other, further limiting the adoption of sustainable finance strategies across the region.

Moreover, limited access to finance remains a critical challenge. Many sustainable projects in Arab countries struggle to attract funding due to perceived risks and a lack of financial infrastructure, particularly for small and medium-sized enterprises (SMEs) (Al Ghunaimi & Awashreh, 2025), and early-stage ventures in sectors such as renewable energy and climate change adaptation (United Nations, 2023; Franczak & Warner, 2024). Without addressing this financing gap, it will be difficult to scale up sustainable projects that are essential for achieving long-term development goals.

This collaborative method allows for the identification of practical, actionable steps for enhancing sustainable finance in the region. The combination of research methods and stakeholder engagement will offer a comprehensive understanding of the challenges and opportunities in advancing sustainable finance across Arab countries. Figure 1 shows present sustainable finance in the Arab region framework (Dushkova & Kuhlicke, 2024).

Figure 1. Sustainable finance framework in the Arab region

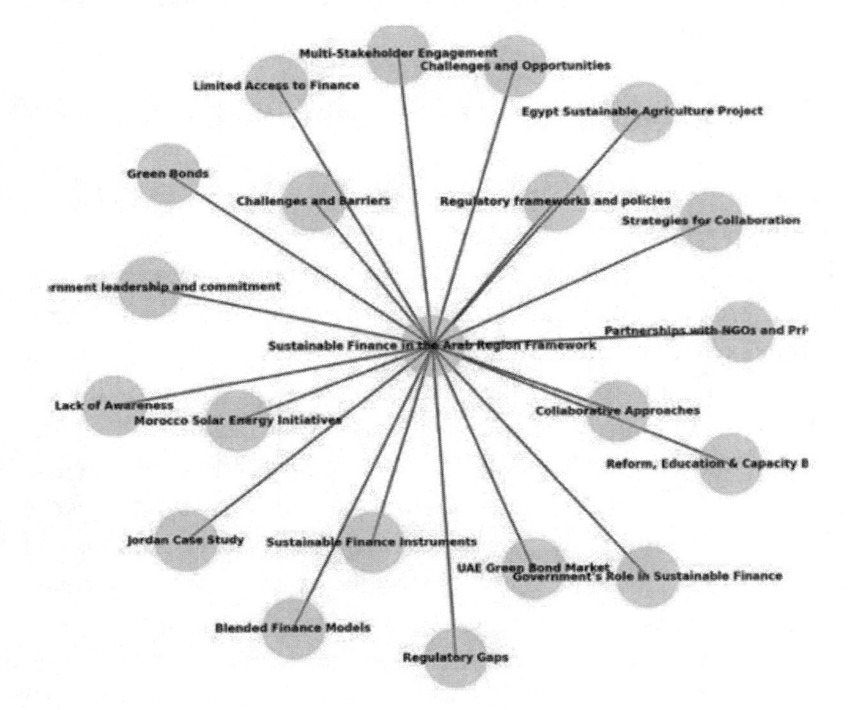

The figure above serves as a comprehensive overview of the various topics and sub-topics within sustainable finance, specifically addressing the roles, instruments, case studies, collaborations, and challenges within the Arab region.

4. RESULTS

This section presents the findings from the analysis, highlighting key aspects of sustainable finance in the Arab region. It covers successful initiatives, collaborative approaches for sustainable development, and the challenges and barriers hindering progress. Additionally, it explores strategies for overcoming these obstacles to advance sustainable finance.

4.1 Successful Sustainable Finance Initiatives

The Arab region is witnessing a rise in successful sustainable finance initiatives that address socio-economic and environmental challenges, driven by collaborations between governments, the private sector, and international partners (Zaidan, Al-Saidi & Hammad, 2019). These initiatives demonstrate the growing recognition of sustainable finance as a vital tool for promoting economic resilience, environmental sustainability, and social equity. By leveraging innovative financing mechanisms and aligning with global sustainability goals, Arab nations are setting benchmarks for green investments and sustainable development.

One prominent example is the UAE, which has taken the lead in green finance through its rapidly expanding green bond market. These financial instruments have been instrumental in funding sustainable infrastructure projects, such as energy-efficient buildings, renewable energy developments, and the Dubai Green Metro Line. The UAE's proactive government role, coupled with strong regulatory frameworks aligned with international standards, has positioned the country as a regional hub for sustainable finance. The success of these efforts has attracted global investors and strengthened the UAE's commitment to achieving its ambitious Net Zero by 2050 initiative (Khiari et al., 2024; Benzaken et al., 2024). Moreover, the UAE's financial sector has introduced sustainability-linked loans and green sukuk (Islamic bonds), expanding financing options for environmentally and socially responsible projects.

Similarly, Egypt has made significant strides in sustainable finance through its National Project for Sustainable Agriculture, which leverages blended finance to enhance agricultural productivity and environmental sustainability. This initiative focuses on attracting private sector investments in water-efficient irrigation systems, climate-resilient crops, and sustainable farming techniques, ensuring long-term food security while mitigating the adverse effects of climate change. International donors, including the World Bank and the Green Climate Fund, play a crucial role in de-risking private investments and providing technical support for implementing sustainable agricultural practices (Kühlert et al., 2024). Additionally, Egypt has issued its first sovereign green bond, which has financed projects in clean transportation, renewable energy, and pollution reduction, demonstrating its commitment to sustainability-driven economic growth.

Morocco's Noor Ouarzazate Solar Complex stands out as a landmark project that showcases the potential of public-private partnerships (PPPs) in driving sustainable finance. As one of the largest concentrated solar power plants in the world, the project has significantly contributed to Morocco's renewable energy capacity and climate mitigation efforts. The initiative combines international loans, government funding, and private sector investment to reduce reliance on fossil fuels and enhance energy security. The successful implementation of this project highlights the importance of long-term planning, policy stability, and cross-sector collaboration in financing large-scale renewable energy projects (Kühlert et al., 2024; Benbba et al., 2024). The Moroccan government has also introduced incentives such as feed-in tariffs and tax benefits to encourage private sector participation in the green energy transition.

In a similar vein, Jordan has addressed its chronic water scarcity challenge through the Disi Water Conveyance Project, which utilizes a build-operate-transfer (BOT) model to finance sustainable water infrastructure. This initiative blends private investment with international support to improve access to clean water and ensure the long-term sustainability of water resources. By securing financing from development banks and private entities, the project has provided a model for financing critical infrastructure in resource-scarce environments (Al-Addous et al., 2023; Marin, 2009). Additionally, Jordan has pioneered new financing mechanisms, such as water impact bonds, to attract investments in water conservation and desalination projects.

Beyond these large-scale projects, several Arab countries are implementing financial innovations to expand sustainable finance. Saudi Arabia, for instance, has launched a sustainability-linked loan framework to encourage businesses to meet ESG targets, while Bahrain has developed green fintech solutions to integrate sustainability into digital finance. Similarly, Tunisia and Lebanon are piloting microfinance programs aimed at supporting small-scale renewable energy and climate adaptation initiatives, ensuring that sustainable finance reaches underserved communities.

These diverse initiatives offer key lessons for policymakers, financial institutions, and businesses across the region. Government leadership, clear policy frameworks, innovative financing mechanisms, community engagement, and regional and international partnerships are essential for scaling sustainable finance and ensuring its long-term success (Gorelick, Cara & Kavoo, 2024; Alhejaili, 2024). Additionally, fostering local expertise in ESG reporting, risk assessment, and impact measurement will further enhance the credibility and effectiveness of sustainable finance initiatives.

Taken together, these cases provide valuable insights for other Arab countries seeking to design and implement sustainable finance strategies. By building on existing successes and adapting proven models to their specific contexts, Arab nations can accelerate their transition towards a more resilient and sustainable economic future. With the right mix of policy support, financial innovation, and cross-sector collaboration, the region has the potential to emerge as a global leader in sustainable finance, driving meaningful change for both present and future generations.

4.2 Collaborative Approaches for Sustainable Development

Sustainable development in the Arab region requires close collaboration among governments, the private sector, non-governmental organizations (NGOs), and international institutions to effectively tackle pressing challenges such as climate change, water scarcity, and social inequality (UNDP, 2013). These challenges are complex and interconnected, demanding coordinated efforts that integrate economic, social, and environmental considerations. In this context, multi-stakeholder engagement plays a vital role in mobilizing resources, driving innovation, and developing sustainable solutions. By leveraging the strengths and expertise of various actors, these partnerships create synergies that enhance the effectiveness of sustainable finance initiatives and contribute to long-term development goals.

Governments play a key role in setting policy frameworks, providing regulatory incentives, and ensuring alignment with national development strategies. The private sector, in turn, brings capital, technological innovation, and operational expertise, helping to implement scalable sustainability solutions. NGOs contribute by offering local knowledge, advocating for environmental and social issues, and engaging communities in sustainability initiatives. International institutions provide technical support, funding, and best practices from global experiences, further strengthening local efforts. Together, these stakeholders help overcome resource constraints, reduce investment risks, and ensure that sustainable projects are both financially viable and socially impactful (Reynolds, 2024).

Several successful initiatives in the Arab region illustrate the power of multi-stakeholder collaboration in advancing sustainable finance. In the UAE, the partnership between the government and NGOs in projects like Masdar City exemplifies how cooperation can drive sustainable urban development. Masdar City serves as a model for energy-efficient architecture and green technologies, attracting international investors and technology firms committed to sustainability (Griffiths & Sovacool, 2020; Saradara et al., 2023). The city's innovative design and infrastructure have demonstrated how sustainable cities can reduce carbon footprints while fostering economic growth.

Similarly, Egypt's Integrated Solid Waste Management Program, implemented through collaboration between the Ministry of Environment, private sector stakeholders, and NGOs, has significantly improved waste reduction and recycling efforts. This initiative not only addresses environmental concerns but also creates economic opportunities through job creation in the waste management sector (Ibrahim & Mohamed, 2016). By promoting circular economy principles, the program demonstrates how sustainable finance can support waste management systems that benefit both society and the environment.

In Morocco, the Noor Ouarzazate Solar Complex stands as one of the largest solar power projects globally, financed through a mix of public and private investments. This project exemplifies the role of blended finance, where public funds de-risk private investment in sustainable infrastructure. The success of Noor Ouarzazate underscores the importance of cross-sector collaboration in renewable energy development, enhancing energy security while reducing greenhouse gas emissions (Benbba et al., 2024). The project has also strengthened Morocco's position as a leader in renewable energy, with the potential to inspire similar initiatives across the region.

The effectiveness of these multi-stakeholder partnerships is largely driven by their ability to mobilize resources, encourage knowledge sharing, and foster innovation. Regional platforms, such as the Arab Forum for Sustainable Development, provide crucial opportunities for dialogue, allowing stakeholders to exchange best practices, discuss policy solutions, and enhance regional cooperation (Eweje et al., 2020). Additionally, joint funding mechanisms, such as blended finance, play a critical role in attracting private investment by mitigating risks and ensuring financial sustainability (Jung, 2020). By combining public and private capital, blended finance can unlock significant resources for climate adaptation, infrastructure development, and social impact projects.

Capacity-building initiatives further enhance the long-term success of sustainable finance efforts in the Arab world. Institutions such as the Arab League's Arab Ministerial Council for Environment actively promote sustainability principles by providing training programs, research support, and policy guidance (Clark, Reed & Sunderland, 2018). Equipping policy-makers, financial professionals, and entrepreneurs with the necessary skills to integrate sustainability into investment decisions is essential for scaling sustainable finance.

Moreover, transparency and accountability are critical components of sustainable finance. Governments and financial institutions must establish mechanisms to track financial flows, assess project outcomes, and apply Environmental, Social, and Governance (ESG) criteria to investment decisions. Ensuring that financial resources are directed toward projects with measurable environmental and social benefits will strengthen investor confidence and enhance the credibility of sustainable finance initiatives (Boscia et al., 2019). Furthermore, adopting standardized ESG reporting frameworks and integrating sustainability assessments into financial regulations will help prevent greenwashing and ensure that investments align with national and global sustainability targets (Ezeh et al., 2024).

In summary, multi-stakeholder engagement is fundamental to advancing sustainable finance in the Arab region. By leveraging the strengths of governments, private enterprises, NGOs, and international institutions, countries can develop innovative financial solutions that address complex environmental and socio-economic challenges. Transparent governance mechanisms, knowledge-sharing platforms, and capacity-building initiatives will further accelerate the growth of sustainable finance, ensuring a resilient and sustainable future for the region. With continued collaboration and strategic investment, Arab nations have the potential to become global leaders in sustainable finance, setting an example for other regions navigating similar development challenges (Alhejaili, 2024).

4.3 Challenges and Barriers to Advancing Finance

Advancing sustainable finance in Arab countries is essential for fostering long-term economic, social, and environmental progress. However, several challenges hinder its expansion, including regulatory gaps, lack of awareness, and limited access to finance. Overcoming these obstacles requires a comprehensive approach that integrates policy reform, education, capacity building, and financial innovation (Alhejaili, 2024).

A key challenge is the regulatory gap, as many Arab countries lack clear, standardized policies and frameworks to support sustainable investments. The absence of well-defined regulations creates uncertainty, discouraging investors and limiting the growth of green financial markets (Agrawal et al., 2024). Furthermore, inconsistencies in Environmental, Social, and Governance (ESG) reporting frameworks complicate project assessments and hinder investor confidence. Without standardized ESG disclosure requirements, businesses and financial institutions struggle to align their practices with international sustainability standards. Additionally, underdeveloped regulations for green bonds, sustainability-linked loans, and blended finance mechanisms further restrict capital flows toward sustainable projects, making it difficult to scale sustainable finance initiatives (Agrawal et al., 2024).

Beyond regulatory constraints, a significant barrier is the lack of awareness among key stakeholders, including governments, financial institutions, and businesses. Many stakeholders continue to prioritize short-term financial returns over long-term sustainability, often due to limited knowledge about sustainable finance benefits and mechanisms. Moreover, policymakers and regulatory bodies frequently lack the expertise needed to implement effective policies that support sustainable investments (Rahman et al., 2023). This knowledge gap delays the adoption of best practices and limits the integration of sustainability considerations into national economic strategies.

Limited access to finance further exacerbates these challenges. While global capital flows toward sustainable investments are increasing, many projects in the Arab region struggle to secure funding due to high perceived risks and weak financial infrastructure. Small and medium-sized enterprises (SMEs) focusing on sustainability-driven innovations often face difficulties accessing traditional bank financing. Additionally, early-stage projects require substantial upfront capital, but investors may be deterred by uncertain returns and regulatory ambiguities (Awashreh & Zaabanut, 2025; PwC, 2024). The lack of specialized financial products tailored for green and social enterprises also constrains the growth of sustainable finance in the region.

To address these challenges, Arab governments must implement targeted policy reforms that establish clear regulatory frameworks for sustainable finance. Specifically, harmonizing ESG reporting standards, facilitating green bond issuance, and developing blended finance mechanisms will help build investor confidence and ensure that capital flows toward environmentally and socially responsible projects (Khan et al., 2024). Stronger policy incentives, such as tax benefits for green investments and sustainability-linked financial instruments, can further encourage businesses and financial institutions to integrate ESG principles into their operations.

Education and awareness campaigns are also essential for expanding sustainable finance adoption. These initiatives should target financial institutions, policymakers, businesses, and the general public to enhance understanding of sustainable finance principles, the importance of ESG considerations, and the long-term benefits of green investments. Public-private partnerships can play a crucial role in delivering training programs and disseminating best practices to improve financial literacy in sustainability-related fields (Rahman et al., 2023).

Capacity building is another critical component in advancing sustainable finance. Strengthening the expertise of financial institutions in assessing ESG risks and structuring sustainable investment products will enable the development of more effective financing solutions. Governments should invest in training programs for public sector officials, ensuring they have the skills needed to design and implement sustainability-focused financial policies (Dzinamarira et al., 2024; Martínez-Peláez et al., 2024). Additionally, integrating sustainable finance topics into university curricula and professional training programs will help build a skilled workforce equipped to drive the green economy.

Strengthening the role of development banks in financing sustainable projects is equally important. These institutions can help reduce investment risks by providing guarantees, concessional financing, and technical assistance for sustainability-focused initiatives. By leveraging public funds to attract private sector investments, development banks can accelerate the deployment of capital toward climate-resilient infrastructure, renewable energy, and social impact projects (Taghizadeh-Hesary & Yoshino, 2020).

Public-private partnerships (PPPs) also offer a viable solution for overcoming financing challenges. By pooling resources and sharing risks, PPPs can facilitate large-scale investments in green infrastructure, energy efficiency, and sustainable urban development. Effective collaboration between governments, private investors, and international financial institutions will be crucial in mobilizing the necessary capital to meet sustainability targets (Vassileva, 2022).

In summary, addressing the challenges of regulatory gaps, stakeholder awareness, and limited access to finance requires a coordinated, multi-faceted approach. By implementing policy reforms, enhancing financial literacy, strengthening institutional capacity, and fostering collaboration between public and private actors, Arab countries can unlock the full potential of sustainable finance. These efforts will not only contribute to regional economic resilience but also align the Arab region with global sustainability objectives, paving the way for a more inclusive and environmentally responsible future (Alhejaili, 2024).

4.4 Overcoming Barriers to Sustainable Finance

Advancing sustainable finance in Arab countries faces challenges that hinder the potential of investments for long-term economic, social, and environmental progress (UN, 2024). These barriers, such as regulatory gaps, lack of awareness, and limited access to finance, must be addressed through comprehensive strategies, including policy reform, education, and capacity building (Alhejaili, 2024). One of the primary challenges is the regulatory gap. Many Arab countries lack clear, standardized policies that incentivize sustainable investments, and inconsistent frameworks for Environmental, Social, and Governance (ESG) reporting hinder investor confidence (Agrawal et al., 2024). Additionally, another significant barrier is the lack of awareness of sustainable finance among key stakeholders, including governments, financial institutions, and businesses. Knowledge gaps persist, especially in the private sector, where short-term financial gains are often prioritized over long-term sustainability (GTR, 2024). These issues compound each other, further limiting the adoption of sustainable finance strategies across the region.

Moreover, limited access to finance remains a critical challenge. Many sustainable projects in Arab countries struggle to attract funding due to perceived risks and a lack of financial infrastructure, particularly for small and medium-sized enterprises (SMEs) (Al Ghunaimi & Awashreh, 2025), and early-stage ventures in sectors such as renewable energy and climate change adaptation (United Nations, 2023; Franczak & Warner, 2024). Without addressing this financing gap, it will be difficult to scale up sustainable projects that are essential for achieving long-term development goals.

To advance sustainable finance, policy reform is urgently needed to address regulatory gaps and create an environment conducive to sustainable investments. Governments must develop clear regulations for ESG reporting, green bonds, and blended finance mechanisms (Khan et al., 2024). In parallel, education and awareness campaigns are essential to improving the understanding of sustainable finance, as well as incentivizing businesses to align with global sustainability goals (Rani et al., 2023). Furthermore, capacity building plays a pivotal role, involving training in ESG risk assessment and the structuring of sustainable investment products, which will empower stakeholders to participate more actively in sustainable finance initiatives (Junaedi, 2024).

In addition, strengthening the role of development banks is key to mitigating investment risks and providing financing for sustainable projects. Public-private partnerships (PPPs) can help pool resources and share risks, ensuring that large-scale sustainable projects align with national development goals (Taghizadeh-Hesary & Yoshino, 2020; Vassileva, 2022). By combining public sector resources with private sector innovation, these partnerships can create the financial stability needed for long-term success. In short, addressing these challenges with coordinated efforts can unlock the full potential of sustainable finance in the Arab region. By focusing on policy reform, education, capacity building, and strategic partnerships, Arab countries can drive long-term growth and contribute to global sustainability goals. Ultimately, these efforts will lay the foundation for a more sustainable and resilient future.

5. CONCLUSION

Sustainable finance presents a significant opportunity for the Arab region to accelerate its transition to a more sustainable and resilient future. However, the region faces several key barriers, including regulatory gaps, lack of awareness, and limited access to finance. To overcome these challenges and fully realize the potential of sustainable finance, a multi-pronged approach is necessary, involving comprehensive policy reform, educational initiatives, and capacity building. Governments, financial institutions, businesses, and civil society must work together to create an enabling environment for sustainable investment, ensuring that both the public and private sectors are equipped to tackle environmental, social, and economic challenges effectively. By addressing regulatory gaps, fostering greater awareness of sustainable finance, and improving access to capital, Arab countries can unlock new sources of financing for sustainable development. The region has the potential to become a global leader in sustainable finance, driving innovation and economic growth while contributing to the achievement of global sustainability goals. However, achieving this requires strategic action at all levels, from government policies and institutional frameworks to practical implementation and ongoing research.

In this context, future research is crucial to better understand the long-term impacts of sustainable finance mechanisms, and to explore ways in which regional challenges can be overcome through innovation. Identifying limitations such as gaps in data, resource constraints, or insufficient infrastructure is vital to shaping effective policy recommendations.

The chapter provides few recommendations:

- Strengthen Regulatory Frameworks: Governments should implement standardized regulations for ESG reporting, green bonds, and blended finance mechanisms to boost investor confidence and private sector participation.
- Promote Public-Private Partnerships: Encourage partnerships between the public and private sectors to finance large-scale sustainable projects like renewable energy and climate change mitigation.
- Increase Awareness and Education: Launch campaigns to educate policymakers, financial institutions, businesses, and the public on sustainable finance, integrating sustainability topics into educational curricula.
- Enhance Capacity Building: Invest in training for financial professionals and government bodies to improve expertise in ESG risks and sustainable financial products.
- Foster Access to Finance: Explore new financial instruments like blended finance to enhance capital access for sustainable projects, especially for SMEs and emerging sectors.

Sustainable finance promotes inclusive economic growth, responsible corporate behavior, and improved living standards. It addresses critical issues such as poverty, education, and access to clean energy while fostering resilient communities. In order to achieve these objectives, financial institutions will need to adapt by integrating Environmental, Social, and Governance (ESG) factors into their risk management strategies. Simultaneously, governments must create supportive regulatory environments that encourage sustainable investment practices.

While the potential for sustainable finance is immense, the research also acknowledges certain limitations. For instance, the lack of comprehensive ESG data and the slow pace of regulatory reforms may impede progress. Future research should focus on overcoming these barriers and exploring strategies to improve data availability and regulatory coherence across the region.

In this regard, the article suggests several government policies that could further support sustainable finance. Governments should incentivize sustainable investments through tax benefits, guarantees, and clear ESG reporting guidelines. Furthermore, integrating sustainability into national development policies is essential for fostering collaboration across sectors and achieving long-term sustainability goals.

It is important to recognize that the full integration of sustainable finance into policy may face challenges, such as political resistance or competing national priorities. Addressing these constraints through targeted research will be critical to achieving sustainable finance objectives.

From a practical perspective, implementing sustainable finance requires effective collaboration between the public and private sectors. Financial institutions must develop the infrastructure necessary to assess and manage ESG investments, while companies should align their strategies with broader sustainability goals. Additionally, public sector entities must integrate sustainability into national plans and develop financial products, such as green bonds and ESG funds, that encourage responsible investments. Looking forward, further research is needed to explore various dimensions of sustainable finance, particularly in the Arab context. This includes examining the effectiveness of blended finance mechanisms, the development of ESG standards, and the impact of public-private partnerships. Additionally, research should focus on how emerging technologies, such as blockchain, can improve transparency in sustainable finance and assess its social and environmental effects, particularly its role in poverty reduction and building community resilience.

In short, advancing sustainable finance in the Arab region is crucial for addressing pressing challenges like climate change, social inequality, and economic instability. By following the recommendations outlined, the region can unlock the full potential of sustainable finance and contribute to the global transition toward a low-carbon, inclusive, and resilient economy.

REFERENCES

G20 Sustainable Finance Working Group. (2023). *Sustainable finance report: Volume I.* https://g20sfwg.org/wp-content/uploads/2023/10/Volume-I-G20-India-Final-VF.pdf

Abalkina, A., & Zaytsev, Y. (2021). *Private sector engagement in LDCs: Challenges and gaps.*

Abdel-Meguid, A. M., Dahawy, K., & Shehata, N. (2021). Do Egyptian listed companies support SDGs? Evidence from UNCTAD guidance on core indicators disclosures. *Corporate Governance and Sustainability Review, 5*(2), 73–81. DOI: 10.22495/cgsrv5i2p6

Abo-Khalil, A. G. (2024). Integrating sustainability into higher education: Challenges and opportunities for universities worldwide. *Heliyon, 10*(9), e29946. DOI: 10.1016/j.heliyon.2024.e29946 PMID: 38707336

Agrawal, R., Agrawal, S., Samadhiya, A., Kumar, A., Luthra, S., & Jain, V. (2024). Adoption of green finance and green innovation for achieving circularity: An exploratory review and future directions. *Geoscience Frontiers, 15*(4), 101669. DOI: 10.1016/j.gsf.2023.101669

Al-Addous, M., Bdour, M., Alnaief, M., Rabaiah, S., & Schweimanns, N. (2023). Water resources in Jordan: A review of current challenges and future opportunities. *Water (Basel), 15*(21), 3729. DOI: 10.3390/w15213729

Al Ghunaimi, H., & Awashreh, R. (2025). Enhancing sustainability and competitiveness of self-funded SMEs: The hotel model. *International Journal of Innovative Research and Scientific Studies, 8*(1), 1806–1817. DOI: 10.53894/ijirss.v8i1.4805

Al-Qahtani, S., & Albakjaji, M. (2023). The role of the legal frameworks in attracting foreign investments: The case of Saudi Arabia. *Asian Journal of Economic and Environmental Journal.* Retrieved from https://ajee-journal.com/the-role-of-the-legal-frameworks-in-attracting-foreign-investments-the-case-of-saudi-arabia

Alhammadi, A., Alsyouf, I., Semeraro, C., & Obaideen, K. (2024). The role of Industry 4.0 in advancing sustainability development: A focus review in the United Arab Emirates. *Cleaner Engineering and Technology*, *18*, 100708. DOI: 10.1016/j.clet.2023.100708

Alharbi, F. R., & Csala, D. (2020). GCC countries' renewable energy penetration and the progress of their energy sector projects. *IEEE Access : Practical Innovations, Open Solutions*, *8*, 211986–212002. DOI: 10.1109/ACCESS.2020.3039936

Alhejaili, M. O. (2024). Integrating Climate Change Risks and Sustainability Goals into Saudi Arabia's Financial Regulation: Pathways to Green Finance. *Sustainability (Basel)*, *16*(10), 4159. DOI: 10.3390/su16104159

Awashreh, R. (2025). Leadership strategies for managing change and fostering innovation over private higher education: Academic institutions in Oman. In *Evolving strategies for organizational management and performance evaluation*. IGI Global., DOI: 10.4018/979-8-3373-0149-5.ch018

Awashreh, R., Al-Ghunaimi, H., Saleh, R., & Al-Bahri, M. N. (2024). The impact of leadership roles and strategies on employees' job satisfaction in Oman. *Pakistan Journal of Life and Social Sciences*, *22*(2). Advance online publication. DOI: 10.57239/PJLSS-2024-22.2.00120

Awashreh, R., & Hassiba, A. (2025). Green human resources management for sustainable development in emerging economies. In *Examining green human resources management and nascent entrepreneurship* (pp. 51–78). IGI Global., DOI: 10.4018/979-8-3693-7046-9.ch003

Awashreh, R., & Mohamed, Y. (2024). A comparative study of leadership perceptions in government and education sectors: Insights from the UAE and Oman. *The International Journal of Organizational Diversity*, *24*(2), 109–123. DOI: 10.18848/2328-6261/CGP/v24i02/109-123

Awashreh, R., & Zaabanut, M. (2025). The role of foreign companies operating in Omani freezone in revitalizing small and middle enterprises. In E. AlDhaen, A. Braganza, A. Hamdan, & W. Chen (Eds.), *Business sustainability with artificial intelligence (AI): Challenges and opportunities* (pp. 999–1008). Springer Nature. https://doi.org/DOI: 10.1007/978-3-031-71526-6_89

Benbba, R., Barhdadi, M., Ficarella, A., Manente, G., Romano, M. P., El Hachemi, N., Barhdadi, A., Al-Salaymeh, A., & Outzourhit, A. (2024). Solar energy resource and power generation in Morocco: Current situation, potential, and future perspective. *Resources*, *13*(10), 140. DOI: 10.3390/resources13100140

Benzaken, D., Adam, J. P., Virdin, J., & Voyer, M. (2024). From concept to practice: Financing sustainable blue economy in Small Island Developing States, lessons learnt from the Seychelles experience. *Marine Policy*, *163*, 106072. Advance online publication. DOI: 10.1016/j.marpol.2024.106072

Berthod, O. (2018). Institutional theory of organizations. In Farazmand, A. (Ed.), *Global Encyclopedia of Public Administration, Public Policy, and Governance* (pp. 1–5). Springer., DOI: 10.1007/978-3-319-20928-9_63

Boscia, V., Stefanelli, V., Coluccia, B., & De Leo, F. (2019). The role of finance in environmental protection: A report on regulators' perspective. *Risk Governance and Control: Financial Markets & Institutions*, *9*(4), 30–40. DOI: 10.22495/rgcv9i4p3

Carè, R., & De Lisa, R. (2019). Social impact bonds for a sustainable welfare state: The role of enabling factors. *Sustainability (Basel)*, *11*(10), 2884. DOI: 10.3390/su11102884

Clark, R., Reed, J., & Sunderland, T. (2018). Bridging funding gaps for climate and sustainable development: Pitfalls, progress, and potential of private finance. *Land Use Policy*, *71*, 335–346. DOI: 10.1016/j.landusepol.2017.12.013

Dirie, K. A., Alam, M. M., & Maamor, S. (2023). Islamic social finance for achieving sustainable development goals: A systematic literature review and future research agenda. *International Journal of Ethics and Systems*, *40*(1), 676–698. Advance online publication. DOI: 10.1108/IJOES-12-2022-0317

Dushkova, D., & Kuhlicke, C. (2024). Making co-creation operational: A RECONECT seven-steps-pathway and practical guide for co-creating nature-based solutions. *MethodsX*, *12*, 102495. DOI: 10.1016/j.mex.2023.102495 PMID: 38170128

Dzinamarira, T., Bopoto, E., Mutiro, T., Chitungo, I., & Dzinamarira, T. (2024). Overcoming challenges and thriving as a new NGO – a case for institutional capacity building. *Development in Practice*, *34*(8), 949–953. DOI: 10.1080/09614524.2024.2358325

Elamin, M. O. (2023). Advancing ethical and sustainable economy: Islamic finance solutions for environmental, social, & economic challenges in the digital age. *International Journal of Membrane Science and Technology, 10*(5), 408–429. DOI: 10.15379/ijmst.v10i5.2515

Eweje, G., Sajjad, A., Nath, S. D., & Kobayashi, K. (2020). Multi-stakeholder partnerships: A catalyst to achieve sustainable development goals. *Marketing Intelligence & Planning*. Advance online publication. DOI: 10.1108/MIP-04-2020-0135

Falcone, P. M. (2020). Environmental regulation and green investments: The role of green finance. *International Journal of Green Economics, 14*(2), 159–173. DOI: 10.1504/IJGE.2020.109735

Fu, C., Lu, L., & Pirabi, M. (2023). Advancing green finance: A review of sustainable development. *DESD, 1*(1), 20. DOI: 10.1007/s44265-023-00020-3

Khan, M. I., Bicer, Y., Asif, M., Al-Ansari, T. A., Khan, M., Kurniawan, T. A., & Al-Ghamdi, S. G. (2024). The GCC's path to a sustainable future: Navigating the barriers to the adoption of energy efficiency measures in the built environment. *Energy Conversion and Management, X*, 100636.

Gorelick, J., Cara, E., & Kavoo, G. (2024). The fallacy of green municipal bonds in developing countries. *WORLD (Oakland, Calif.), 5*(4), 929–951. DOI: 10.3390/world5040047

Gorelick, J., Cara, E., & Kavoo, G. (2024). The fallacy of green municipal bonds in developing countries. *WORLD (Oakland, Calif.), 5*(4), 929–951. DOI: 10.3390/world5040047

Griffiths, S., & Sovacool, B. K. (2020). Rethinking the future low-carbon city: Carbon neutrality, green design, and sustainability tensions in the making of Masdar City. *Energy Research & Social Science, 62*, 101368. DOI: 10.1016/j.erss.2019.101368

Griffiths, S., & Sovacool, B. K. (2020). Rethinking the future low-carbon city: Carbon neutrality, green design, and sustainability tensions in the making of Masdar City. *Energy Research & Social Science, 62*, 101368. DOI: 10.1016/j.erss.2019.101368

GTR. (2024, April 22). How sustainability is shaping trade finance in the Middle East. *Global Trade Review*. https://www.gtreview.com/magazine/the-esg-trade-issue-2024/how-sustainability-is-shaping-trade-finance-in-the-middle-east/

GTR. (2024, April 22). How sustainability is shaping trade finance in the Middle East. *Global Trade Review*. https://www.gtreview.com/magazine/the-esg-trade-issue-2024/how-sustainability-is-shaping-trade-finance-in-the-middle-east/

Hodge, G., & Greve, C. (2007). Public–private partnerships: An international performance review. *Public Administration Review*, *67*(3), 545–558. DOI: 10.1111/j.1540-6210.2007.00736.x

Iao-Jörgensen, J. (2024). Networking in action: Taking collaborative capacity development seriously for disaster risk management. *Progress in Disaster Science*, *21*, 100311. DOI: 10.1016/j.pdisas.2024.100311

Ibrahim, M., & Mohamed, N. A. E. M. (2016). Towards sustainable management of solid waste in Egypt. *Procedia Environmental Sciences*, *34*, 336–347. DOI: 10.1016/j.proenv.2016.04.030

Ikevuje, A. H. U., Anaba, D. C., & Iheanyichukwu, U. T.Augusta Heavens IkevujeDavid Chinalu AnabaUche Thankgod Iheanyichukwu. (2024). Exploring sustainable finance mechanisms for green energy transition: A comprehensive review and analysis. *Finance & Accounting Research Journal*, *6*(7), 1224–1247. DOI: 10.51594/farj.v6i7.1314

Jung, H. (2020). Development finance, blended finance, and insurance. *International Trade. Politics and Development*, *4*(1), 47–60. DOI: 10.1108/ITPD-12-2019-0011

Kamalov, F., Santandreu Calonge, D., & Gurrib, I. (2023). New era of artificial intelligence in education: Towards a sustainable multifaceted revolution. *Sustainability (Basel)*, *15*(16), 12451. DOI: 10.3390/su151612451

Khiari, W., Ben Flah, I., Lajmi, A., & Bouhleli, F. (2024). The stock market reaction to green bond issuance: A study based on a multidimensional scaling approach. *Journal of Risk and Financial Management*, *17*(9), 408. Advance online publication. DOI: 10.3390/jrfm17090408

Kühlert, M., Klingen, J., Gröne, K., Hennes, L., Terrapon-Pfaff, J. C., & Jamea, E. M. (2024, February). Pathways towards a green economy in Egypt: Final report. Wuppertal Institute for Climate, Environment and Energy. https://epub.wupperinst.org/frontdoor/deliver/index/docId/8534/file/8534_Green_Economy_Egypt.pdf

Marin, P. (2009). Public-private partnerships for urban water utilities. *World Bank.*_https://ppp.worldbank.org/public-private partnership/sites/ppp.worldbank.org/files/documents/FINAL-PPPsforUrbanWaterUtilities-PhMarin.pdf

Martínez-Peláez, R., Ochoa-Brust, A., Rivera, S., Félix, V. G., Ostos, R., Brito, H., Félix, R. A., & Mena, L. J. (2023). Role of digital transformation for achieving sustainability: Mediated role of stakeholders, key capabilities, and technology. *Sustainability (Basel)*, *15*(14), 11221. DOI: 10.3390/su151411221

Mertzanis, C., & Tebourbi, I. (2024). Geopolitical risk and global green bond market growth. *European Financial Management*. Advance online publication. DOI: 10.1111/eufm.12484

OECD. (2020). *Blended finance in the least developed countries 2020: Supporting a resilient COVID-19 recovery*. Organisation for Economic Co-operation and Development.

Polzin, F., Egli, F., Steffen, B., & Schmidt, T. S. (2019). How do policies mobilize private finance for renewable energy?—A systematic review with an investor perspective. *Applied Energy*, *236*, 1249–1268. DOI: 10.1016/j.apenergy.2018.11.098

Porretta, P., & Benassi, A. (2021). Sustainable vs. not sustainable cooperative banks business model: The case of GBCI and the authority view. *Risk Governance and Control: Financial Markets & Institutions*, *11*(1), 33–48. DOI: 10.22495/rgcv11i1p3

PwC. (2024, June 5). *Sustainability in the Middle East 2024*. PwC. https://www.pwc.com/m1/en/sustainability/insights/sustainability-in-the-middle-east-2024.html

Rahman, M. H., Rahman, J., Tanchangya, T., & Esquivias, M. A. (2023). Green banking initiatives and sustainability: A comparative analysis between Bangladesh and India. *Research in Globalization*, *7*, 100184. DOI: 10.1016/j.resglo.2023.100184

Raimi, L., Abdur-Rauf, I. A., & Ashafa, S. A. (2024). Does Islamic sustainable finance support sustainable development goals to avert financial risk in the management of Islamic finance products? A critical literature review. *Journal of Risk and Financial Management*, *17*(6), 236. DOI: 10.3390/jrfm17060236

Rani, A. M., Dariah, A. R., Al Madhoun, W., & Srisusilawati, P. (2023). Awareness of sustainable finance development in the world from a stakeholder perspective. *International Journal of Management and Sustainability*, *12*(3), 323–336. DOI: 10.18488/11.v12i3.3428

Raza, A., Alavi, A. B., & Asif, L. (2024). Sustainability and financial performance in the banking industry of the United Arab Emirates. *Discover Sustainability*, *5*(1), 223. DOI: 10.1007/s43621-024-00414-z

Reynolds, S. (2024). Stakeholder engagement and its impact on supply chain sustainability in the context of renewable energy. *Preprints*. https://doi.org/ DOI: 10.20944/preprints202406.0080.v1

Saradara, S. M., Khalfan, M. M. A., Rauf, A., & Qureshi, R. (2023). On the path towards sustainable construction—the case of the United Arab Emirates: A review. *Sustainability (Basel)*, *15*(19), 14652. DOI: 10.3390/su151914652

Taghizadeh-Hesary, F., & Yoshino, N. (2020). Sustainable solutions for green financing and investment in renewable energy projects. *Energies*, *13*(4), 788. DOI: 10.3390/en13040788

The World Bank Group. (2020). Mobilizing private finance for nature. Retrieved from https://thedocs.worldbank.org/en/doc/916781601304630850-0120022020/original/FinanceforNature28Sepwebversion.pdf

Ukoba, K., Yoro, K. O., Eterigho-Ikelegbe, O., Ibegbulam, C., & Jen, T.-C. (2024). Adaptation of solar energy in the Global South: Prospects, challenges and opportunities. *Heliyon*, *10*(7), e28009. DOI: 10.1016/j.heliyon.2024.e28009 PMID: 38560131

UNEP FI. (2020). Draft for consultation: Promoting sustainable finance & climate finance in the Arab region. United Nations Environment Programme Finance Initiative. https://www.unepfi.org/wordpress/wp-content/uploads/2020/10/ConsultationDraft-Promoting-Sustainable-Finance-in-the-Arab-Region.pdf

Unger, C., & Thielges, S. (2021). Preparing the playing field: Climate club governance of the G20, Climate and Clean Air Coalition, and Under2 Coalition. *Climatic Change*, *167*(3-4), 41. DOI: 10.1007/s10584-021-03189-8

United Nations. (2018). Towards Saudi Arabia's sustainable tomorrow: Kingdom of Saudi Arabia, UN High-Level Political Forum 2018: "Transformation towards sustainable and resilient societies". https://sustainabledevelopment.un.org/content/documents/20230SDGs_English_Report972018_FINAL.pdf

United Nations. (2023). The state of Arab cities report 2022. United Nations Development Programme (UNDP). https://www.undp.org/sites/g/files/zskgke326/files/2024-05/the_state_of_arab_cities_report_2022_final_eng.pdf

United Nations. (2024). The Sustainable Development Goals Report 2024. United Nations. https://unstats.un.org/sdgs/report/2024/The-Sustainable-Development-Goals-Report-2024.pdf

Vassileva, A. (2022). Green public-private partnerships (PPPs) as an instrument for sustainable development. *Journal of World Economy: Transformations & Transitions*, *2*(5), 1–18. DOI: 10.52459/jowett25221122

Weber, O. (2018). The financial sector and the SDGs: Interconnections and future directions (CIGI Book chapters No. 201). Centre for International Governance Innovation. https://www.cigionline.org/static/documents/documents/Book chapter%20No.201web.pdf

World Bank Group. (2017, March). Morocco: Noor Ouarzazate - Concentrated solar power complex. Multilateral Development Banks' Collaboration: Infrastructure Investment Project Briefs. https://ppp.worldbank.org/public-private-partnership/sites/ppp.worldbank.org/files/2022-02/MoroccoNoorQuarzazateSolar_WBG_AfDB_EIB.pdf

Zaidan, E., Al-Saidi, M., & Hammad, S. H. (2019). Sustainable development in the Arab world – Is the Gulf Cooperation Council (GCC) region fit for the challenge? *Development in Practice*, *29*(5), 670–681. DOI: 10.1080/09614524.2019.1628922

Zamiri, M., & Esmaeili, A. (2024). Methods and technologies for supporting knowledge sharing within learning communities: A systematic literature review. *Administrative Sciences*, *14*(1), 17. DOI: 10.3390/admsci14010017

Chapter 9
Quantitative Modelling of Shariah Principles of Islamic Transactions:
Empirical Validation in Islamic and Conventional Fundings

Mohamed Gassouma
https://orcid.org/0000-0002-0932-9326
Department of Islamic Law, Economics, and Finance, Higher Institute of Theology of Tunis, Tunisia

ABSTRACT

This paper aims to assess the importance of adhering to Sharia principles in both conventional and Islamic fundings. We assumed that these principles apply equally to both Islamic and conventional fundings. We also assume that Sharia principles is not confined to a specific space or time but is universal to all financial establishment. Empirically, we attempted to model these principles using ratios, thresholds, and econometric models. The principles used in this study include: the value of labor, the principle of equality between goods and money "Ghonm to Ghorm", the prohibition of fraud and manipulation, the prohibition of currency speculation "Riba", and the prohibition of arbitrage. The results showed that the key principles leading to non-speculative profit such as the prohibition of currency speculation, equality between money and goods, and the importance of labor are better upheld in conventional financing than in Islamic financing. However, the Islamic financing system

DOI: 10.4018/979-8-3693-8079-6.ch009

demonstrated superiority in market-related factors, such as the prohibition of fraud and manipulation.

1. INTRODUCTION

The fundamental principle of Islamic financial Transactions is that banking and market financing must have tangible counterparts. By "tangible," we refer to all goods and services that generate profit or loss. Profit and loss thus become essential components in determining the value of tangible assets. Consequently, financial value derives its legitimacy from the value of tangible assets, making money an equivalent to real assets. This principle aligns with the balance between the monetary and real economies, known in Islamic jurisprudence as the principle of "Al-Ghunm bil-Ghurm" (profit comes with liability). (Mashour, 1991).

Profit (through goods and services) and liability (through money) can only be realized if every cash flow has a corresponding tangible counterpart. Every increase in money must originate from profits on goods, ensuring that the value of goods rises in proportion to the increase in money. If this principle is not followed, an imbalance occurs: either an excess of goods without sufficient money to purchase them or an overabundance of money without available goods.

This financial surplus relative to goods leads to inflation and price increases, a phenomenon known in Islamic jurisprudence as "Al-Ghulla", which results in social inequality and declining purchasing power. (Benhamed and Gassouma, 2023). This excess of money is considered "Riba-based money", meaning wealth acquired outside of trade transactions. In other words, it is the money that increases without legitimate justification, either through fraudulent means or through loans not invested in tangible assets as investment and consumption but used to cover debts and to arbitrage operations.. All such funds fall under "virtual Riba-based money."

This virtual money is generated through trade over time not backed by tangible goods but rather acquired through delayed debt repayments or speculative price increases. The value of goods should rise with the rising of money value corresponding to the labor creation. (Adesina, K.S, 2020).

Thus, the second principle is labor and the value of labor. Work is the only means to create added value in goods and services. Through labor, real profits are generated via production and the transformation of raw materials into finished products. (The added value created through labor is then distributed among employees and partners. In the Islamic theory, Employees fall into two categories: households and state. The state is considered as an indirect worker because it provides infrastructure and services to businesses and remunerated through taxes collection.

Through labor, financial growth becomes real and productive, as it is reinvested into goods and services, creating an added value within the economic cycle. In the absence of labor, production does not occur, and non-productive profits emerge, violating the principle of "Al-Ghunm bil-Ghurm." This added value is distributed as wages to workers, taxes to state, and profit shares to partners.

The third principles is the hoarding preventing and the encouraging spending in investment and consumption. This principle is necessary to avoid speculation and no-tangible investment and consequently ensuring the "Ghunm bil Ghurm" principle. (Gassouma, 2023.a)

The Deposited funds, even in conventional or Islamic banks, are not considered hoarded wealth because deposits are reinvested into the economy through investment loans, consumer loans, partnerships, Murabaha, and other financing instruments. Hoarded wealth is money that remains idle and uninvested.

These last principles can be insured by the Zakat payment and the illicit money given to economy (as poor householders, public and private associations…) through the income's purification process.

So, Islam emphasizes the obligation of Zakat on incomes and on assets after deducting expenses, wages, taxes, and profit shares. The remaining liquidity and assets that exceed the required threshold for one year must be partially distributed to the needy.

Thereby, Zakat can play the role of two functions: preventing wealth hoarding and ensuring wealth circulation. To avoid paying a higher amount of Zaket, households, try to invest their fund in economy and receiving most income, thereby spending and not holding money. This is the first utilities of Zaket called the "Maqasad Shariah of Zaket". Secondly, when households, distribute Zakat to those who deserve it, this allows a well succession of funds and a best fair repartition. (Khan and Mirakhor, 1989).

To ensure compliance with these principles, a stable and transparent financial environment is essential. To ensure this, it is evident to refer to the most critical principle consisting to providing full disclosure on financing and sales transactions and preventing fraud and manipulation. Deceptive practices such as tax evasion, financial misreporting, and price manipulation distort the true value of goods, leading to virtual money creation. One of the key theories addressing financial fraud is the "Accruals Manipulation Theory."

Unfair competition also involves asymmetric information and artificial demand creation, which can drive prices above their equilibrium level, leading to speculative, virtual wealth. In Islamic jurisprudence, this practice is known as "Najsh" (price manipulation). Similarly, withholding information from sellers, monopolizing trade, and restricting sales channels artificially inflates prices, creating Riba-based virtual money. This practice is referred to in Islamic law as "Bai' Al-Hadir lil-Badi" (local seller exploiting external buyer). (Gassouma, 2022)

These factors all operate within a risk environment. Every investment is subject to risk and potential loss. Investors and financiers must mitigate and account for risks, but they cannot eliminate them entirely. However, some risks are avoidable and must be prevented. Avoidable risks that are not managed fall called in Islamic jurisprudence by "Gharar" (excessive uncertainty not able to be recovered). One of the causes of Gharar is market ignorance due to information asymmetry.

Ultimately, these distortions result in a mismatch between the price of goods and the value of money, creating Riba-based wealth that harms the economy and causes financial and social instability.

This paper seeks to answer the following question:

To what extent do Islamic and conventional financing adhere to Sharia-compliant principles?

The legitimacy of a bank does not come from its structure alone but also from the financial behaviour of its clients. If a client receives financing without producing goods or generating profit, or if they manipulate the market to set an investment price inconsistent with their financing terms, they contribute to the creation of virtual money, even if they use Islamic contracts like Murabaha or Mudaraba. (Gassouma et al, 2022.a)

Thus, an Islamic financial product is only truly Sharia-compliant when applied correctly by its recipient, not just its financer. Accordingly, this study aims to assess the legitimacy of institutions receiving Islamic financing to determine the true compliance of the financing itself.

2. DESIGN EMPIRICAL MODEL

In this experimental section, we introduce a model based on Principal Component Analysis (PCA) to assess the contribution of each Shariah compliance factor to the overall Islamic legitimacy of financing. Each compliance factor is conceptualized as a variable and integrated into the model to determine its relative weight in evaluating compliance.

2.1 Sample

The study sample consists of a set of Tunisian Commercial institutions that received bank financing, divided into:

- Institutions that obtained Islamic financing
- Institutions that received conventional financing

The study period spans from 2000 to 2022. Data is sourced from financial statements and annexes available in the Tunis Stock Exchange database.

2.2 Variables

1. The Principle of "Al-Ghunm bil-Ghurm" (Money-goods equilibrium)

This principle ensures that profits are justified by financial liability. The compliance variable is calculated as:

$$\text{Al-Ghunm bil-Ghurm} = \left(\frac{\textbf{Fixed Assets}}{\textbf{Equity + Non − Current Liabilities}}\right) - 1$$

A value of 1 indicates full adherence to the principle, meaning all financial liabilities are backed by tangible assets. A small value of this indicator indicates a legitimacy in the funding.

2. The Principle of Labor

Labor is the only factor that creates real wealth and ensures non-Riba profits for banks. The compliance variable is defined as: (Adesina, 2020)

$$\mathbf{LABOR} = \frac{\mathbf{Revenues - (Operating\ Expenses - Employment\ Costs)}}{\mathbf{Employment\ Costs}}$$

The higher this ratio, the greater the role of labor in wealth generation, ensuring compliance with Islamic finance principles.

3. Prohibition of Hoarding & Encouragement of Spending Money

Islamic finance discourages excessive liquidity hoarding, using one-third as the maximum liquidity threshold based on Islamic jurisprudential guidelines. The compliance variable is: (Gassouma, 2022)

$$Hoarding\ money = \left(\frac{Net\ cash}{Total\ Assets}\right) - 0.33$$

A value exceeding **0.33** indicates excessive liquidity hoarding, deviating from Shariah principles.

4. Prohibition of Monopoly

Monopoly disrupts market equilibrium. To measure monopoly tendencies, we set a one-third threshold for inventory relative to total assets. The compliance variable is: (Gassouma, 2023)

$$Monopoly = \left(\frac{Inventory}{Total\ Assets}\right) - 0.33$$

A higher value suggests an excessive concentration of inventory, which could indicate market manipulation.

5. Prohibition of Fraud

Fraud results in financial distortions and economic losses. To measure fraud, we adopt the Kothari et al. (2005) accrual-based model, which evaluates earnings manipulation:

$$\frac{ACT_{i,t}}{TA_{i,t-1}} = \alpha 0 \times \frac{1}{TA_{i,t}} + \alpha 1 \times \frac{FA_{i,t}}{TA_{i,t}} + \alpha 2 \times \frac{\left(\Delta Turnover_{i,t} - \Delta CUD_{i,t}\right)}{TA_{i,t}} + \alpha 3 \times \frac{NI_{i,t-1}}{TA_{i,t-1}}$$

Where:

- **ACT** = Fraud value = Net Income - Cash Flows
- **TA** = Total Assets
- **FA** = Fixed Assets
- **Turnover** = Revenues
- **CUD** = Customer Receivables
- **NI** = Net Income

After linear adjustments and coefficient estimation, we extract the error margin for each financing, representing the extent of fraudulent activities.

6. Prohibition of Riba (Interest-Based Transactions)

To eliminate Riba, banks must apply profit-based financing instead of interest-based returns. The compliance variable is calculated as: (Gassouma, 2022.a)

$$\textbf{Applied Interest Rate} = \frac{\textbf{Intesrest Paid}}{\textbf{Debt}}$$

A theoretical profit rate based on asset performance is computed as:

$$\textbf{Theorical Intesrest rate} = \left(\frac{\textbf{Accounting Profit}}{\textbf{Total Fixed Assets}}\right) \times \left(\frac{\textbf{Debt Value}}{\textbf{Total Assets}}\right)$$

The greater the gap between the applied interest rate and the theoretical profit rate, the more likely the bank relies on Riba-based returns rather than asset-generated profits.

2.3 Principal Component Analysis Model

Table 1. Principal component analysis

Variables Legitimacy	Legitimacy of Islamic Finance	Legitimacy of Conventional Finance
Money-Goods equilibrium	0.1154*	0.2924*
Hoarding Money	-0.3454**	-0.2791**
Monopoly	-0.3351*	-0.0361*
Labor	0.1891***	0.4082***
Fraud	-0.5128*	-0.3888*
Usury (Al-Riba)	-0.0321**	-0.2522**

*significance at level 10% ** significance level at 5% *** significance level at 1%

3. RESULTS

We observe that adhering to all Sharia principles inevitably leads to the legitimacy of financing, whether in Islamic or conventional financial institutions, albeit with varying degrees between the two banking systems.

We conclude that the contribution of the "Al-Ghunm bil Ghurm" (Money-Goods equilibrum) principle to the legitimacy of conventional financing is higher than its contribution to Islamic financing. Conventional financing maintains a better balance between financing and tangible assets than Islamic financing. Despite the emphasis of Islamic banks on using tangible assets, particularly through Murabaha, this balance is more evident in conventional banks financing. Therefore, we conclude that although Tunisian Islamic banks are diligent in structuring contracts before granting financing, they often overlook post-financing monitoring to ensure a balance between the financed assets and the financing itself. This imbalance can lead to the creation of virtual money derived from a lack of equilibrium. Specifically, the return on assets may be lower than the financing returns, resulting in unreal profits for the bank, which could be categorized as Riba-based earnings.

Regarding the prohibition of hoarding, both banking systems adhere to this principle to ensure the legitimacy of transactions. Islamic banks do so for religious and financial reasons, while conventional banks do so for purely financial motives. The level of compliance with this principle is relatively similar in both systems. Hoarding among borrowers leads to the accumulation of idle funds that are not reinvested in the economic cycle. Consequently, hoarding can reduce overall profitability and create a mismatch between financing returns and asset-generated profits.

As for monopolization, Islamic banks comply with this principle, while conventional banks do not fully adhere to it. Islamic banks gain legitimacy by extending financing to non-monopolists, whereas conventional banks show little concern for this aspect. Granting financing to non-monopolists in Islamic banking helps generate real profits that align with financing returns, ensuring that the bank's earnings originate from the productive assets financed.

Labor is one of the fundamental pillars that contribute to balance and real profitability. However, this principle is more evident in conventional financing than in Islamic financing. In conventional banking, the contribution of work to financing legitimacy is 0.4, compared to 0.18 in Islamic banking. This means conventional banks respect this principle twice as much as Islamic banks. The lack of emphasis on work in Islamic financing may result in non-productive financed assets, preventing the creation of added value that should be distributed among salaries and profit shares.

Fraud prevention is a critical issue that affects price equilibrium. Islamic banks place greater importance on ensuring financing legitimacy by avoiding loans to those with financial misconduct, with a 0.5 emphasis on fraud prevention. Conventional banks, while not far behind, allocate 0.3 to this principle, not necessarily for Sharia compliance but rather to maintain the legitimacy of their financing operations.

At the bottom of the Sharia compliance hierarchy, as outlined in the theoretical framework, lies the prohibition of Riba (usury). This principle is a consequence of failing to the rest of principles. Contrary to expectations, Islamic financing does not place significant weight on ensuring compliance with Riba prohibition. In contrast, conventional banks respect this principle at a level of 0.25, which is relatively high. However, adherence to Riba prohibition in conventional financing is not based on religious legitimacy but rather on creditworthiness considerations. Conventional banks aim to ensure continuous repayment by maintaining profitability, offering loans with interest rates that align closely with the returns on financed assets. Their pursuit of financial stability is framed as legitimacy, even if not intentionally sought for religious purposes.

Notably, Islamic financing lacks strong adherence to Riba prohibition due to its failure to respect Al-Ghunm bil Ghurm and labor-based earnings. These two elements are the fundamental pillars for generating real profits in banks, ensuring that earnings are tied to tangible and productive assets rather than virtual financial instruments detached from real economic balances.

4. CONCLUSION

This study concludes that Islamic financing surpasses conventional financing in terms of compliance with fraud prevention, anti-monopoly, and anti-hoarding principles. Islamic banks strive to provide transparent and stable competitive conditions for their beneficiaries. Conventional banks, while also adhering to these principles to some extent, place little emphasis on anti-monopoly practices.

On the other hand, conventional banks adhere more closely to the principles that lead to Riba prohibition than Islamic banks. While Islamic banks structure their contracts in line with jurisprudential and legal Sharia standards, the economic consequences of these contracts often fail to achieve the intended objectives of Al-Ghunm bil Ghurm and labor-based earnings. Islamic financing tends to align with asset and monetary balance in theory but fails to maintain this balance after the contract is executed, which contradicts Sharia objectives.

In Islamic banking, the bank's responsibility does not end once the financing is granted. it begins at that point. Banks, especially Islamic banks, are partners in investment, not mere regulators. Although Murabaha is a sales contract, it does not exempt banks from responsibility during the repayment period, as it is a long-term agreement rather than an instantaneous transaction. Moreover, Murabaha is based on Mudaraba, and Mudaraba involves diminishing partnerships.

The general conclusion of this study is that the legitimacy of financial transactions does not have a fixed percentage or a commercial label. it derives from its objectives and outcomes. Legitimacy can exist in both conventional and Islamic banking, which is a testament to the comprehensive nature of Islamic principles.

REFERENCES

Adesina, K. S. (2020). How diversification affects bank performance: The role of human capital. *Economic Modelling*.

Askari, H., & Mirakhor, A. (2017). *Ideal Islamic Economy*. Book.

Benhamed, A., & Gassouma, M. S. (2023). Preventing Oil Shock Inflation: Sustainable Development Mechanisms vs. Islamic Mechanisms. *Sustainability (Basel), 15*(12), 2. DOI: 10.3390/su15129837

Gassouma, M. S. (2022). Modeling and Conceptualization of an Islamic Financial System, 5th International Scientific Conference Islam and Contemporary Issues, Turkey – Antalya, 2022, Euro-Arab Organization for Water and Desert & University of Zitouna.

Gassouma, M. S. (2023). Applications of an Alternative Islamic Profit Rate to Interest Rates: Examining Its Impact on Investment and Inflation, *2nd International Conference on Islamic Finance*, Tunisia – Hammamet, University of Zitouna.

Gassouma, M. S. (2023.a). Empirical Modeling of Sharia Compliance in Islamic Financial Transactions: A Case Study on the Alignment of Islamic and Conventional Finance with Sharia Standards, 6th International Scientific Conference Islam and Contemporary Issues, Turkey – Antalya, 2023, Euro-Arab Organization for Water and Desert & University of Zitouna.

Gassouma, M. S., Benahmed, A., & Montasser, G. (2022). (in press). a). Investigating similarities between Islamic and conventional banks in GCC countries: A dynamic time warping approach. *International Journal of Islamic and Middle Eastern Finance and Management.*

Khan, M. S., & Mirakhor, M. (1989). The financial system of monetary policy in an Islamic Economy. *Islamic Economic, Vol, 1*(1), 39–57. DOI: 10.4197/islec.1-1.2

Kothari, S. P., Leone, A. J., & Wasley, C. E. (2005). Performance matched discretionary accrual measures. *Journal of Accounting and Economics, 39*(1), 163–197. DOI: 10.1016/j.jacceco.2004.11.002

Mashhour, A. (1991). *Investment in Islamic Economics, Madbouly Library*. Arabic.

Siddiqi, M. N. (1982). Islamic Approaches to Money, Banking and Monetary Policy: A Review. In Ariff, M. (Ed.), *Monetary and Fiscal Economics of Islam*. International Centre for Research in Islamic Economics.

Compilation of References

Abalkina, A., & Zaytsev, Y. (2021). *Private sector engagement in LDCs: Challenges and gaps.*

Abdel-Meguid, A. M., Dahawy, K., & Shehata, N. (2021). Do Egyptian listed companies support SDGs? Evidence from UNCTAD guidance on core indicators disclosures. *Corporate Governance and Sustainability Review, 5*(2), 73–81. DOI: 10.22495/cgsrv5i2p6

Abedifar, P., Ebrahim, S. M., Molyneux, P., & Tarazi, A. (2016). Islamic banking and finance: Recent empirical literature and directions for future research. In *A Collection of Reviews on Savings and Wealth Accumulation* (pp. 59–91). Wiley. DOI: 10.1002/9781119158424.ch4

Abojeib, M., & Habib, F. (2019). Blockchain for Islamic Social Responsibility Institutions. In *Advances in finance, accounting, and economics book series* (pp. 221–240). DOI: 10.4018/978-1-5225-7805-5.ch010

Abojeib, M., & Habib, F. (2021). Blockchain for Islamic social responsibility institutions. In *Research Anthology on Blockchain Technology in Business, Healthcare, Education, and Government* (pp. 1114–1128). IGI Global. DOI: 10.4018/978-1-7998-5351-0.ch061

Abo-Khalil, A. G. (2024). Integrating sustainability into higher education: Challenges and opportunities for universities worldwide. *Heliyon, 10*(9), e29946. DOI: 10.1016/j.heliyon.2024.e29946 PMID: 38707336

Adesina, K. S. (2020). How diversification affects bank performance: The role of human capital. *Economic Modelling.*

African Development Bank. (2011). *Islamic Banking and Finance in North Africa Past Development and Future Potential.*

Agger, I., & Jensen, S. (1996). *Trauma and Healing Under State Terrorism.* ZEB Books.

Agievich, V. (2014). Mathematical model and multi-criteria analysis of designing large-scale enterprise roadmap. PhD thesis on the specialty 05.13.18 – Mathematical modelling, numerical methods and complexes of programs.

Agrawal, R., Agrawal, S., Samadhiya, A., Kumar, A., Luthra, S., & Jain, V. (2024). Adoption of green finance and green innovation for achieving circularity: An exploratory review and future directions. *Geoscience Frontiers, 15*(4), 101669. DOI: 10.1016/j.gsf.2023.101669

Ahmad, Z. (1984). Concept and Models of Islamic Banking: An Assessment," paper presented at the Seminar on Islamization of Banking, Karachi.

Ahmad, I., Nasir, M., & Mokhtar, N. (2021). *Journal of Islamic Accounting and Business Research, 12*(2), 310–327.

Aktürk, B. (2021). *İslami finansta finansal teknoloji (FinTek) ve FinTek'in katılım bankaları uygulamaları.* https://openaccess.marmara.edu.tr/items/e6649a77-5d1f-42fa-834a-d34c7af2d989

Aktürk, B. (2024). İslami Fintek:Yenilikçi Finans Çözümleri in *İKAM ARAŞTIRMA RAPORLARI* Vol. 32, İLKE Yayın. https://ikam.org.tr/images/fintek_raporu/ikam_arastirma_raporu_32.pdf

Al Ghunaimi, H., & Awashreh, R. (2025). Enhancing sustainability and competitiveness of self-funded SMEs: The hotel model. *International Journal of Innovative Research and Scientific Studies, 8*(1), 1806–1817. DOI: 10.53894/ijirss.v8i1.4805

Al-Addous, M., Bdour, M., Alnaief, M., Rabaiah, S., & Schweimanns, N. (2023). Water resources in Jordan: A review of current challenges and future opportunities. *Water (Basel), 15*(21), 3729. DOI: 10.3390/w15213729

Alam, N., Gupta, L., & Zameni, A. (2019). Fintech and Islamic Finance. In *Springer eBooks.* DOI: 10.1007/978-3-030-24666-2

Alam, N., Gupta, L., & Zameni, A. (2021). *Fintech ve İslami Finans; Dijitalleşme, Kalkınma ve Yenilikçi Yıkım*. Albaraka Yayınları.

Albawaba Business. (2014). Why Islamic Finance is failing in Lebanon. Albawaba Business. https://www.albawaba.com/business/lebanon-islamic-finance-545264

Alderman, L. (2019). French Court Fines UBS $4.2 Billion for Helping Clients Evade Taxes. The New York Times. USA. Retrieved from https://www.nytimes.com/2019/02/20/business/ubs -france-tax-evasion.html

Al-Eitan, G. N., Alkhazaleh, A. M., Alkazali, A. S., & Al-Own, B. (2021). The Internal and External Determinants of the Performance of Jordanian Islamic Banks: A Panel Data Analysis. *Asian Economic and Financial Review*, *11*(8), 644–657. DOI: 10.18488/journal.aefr.2021.118.644.657

Alhammadi, A., Alsyouf, I., Semeraro, C., & Obaideen, K. (2024). The role of Industry 4.0 in advancing sustainability development: A focus review in the United Arab Emirates. *Cleaner Engineering and Technology*, *18*, 100708. DOI: 10.1016/j.clet.2023.100708

Alharbi, F. R., & Csala, D. (2020). GCC countries' renewable energy penetration and the progress of their energy sector projects. *IEEE Access : Practical Innovations, Open Solutions*, *8*, 211986–212002. DOI: 10.1109/ACCESS.2020.3039936

Alhejaili, M. O. (2024). Integrating Climate Change Risks and Sustainability Goals into Saudi Arabia's Financial Regulation: Pathways to Green Finance. *Sustainability (Basel)*, *16*(10), 4159. DOI: 10.3390/su16104159

Ali, S. M., & Abbas, G. (2022). *International Journal of Islamic and Middle Eastern Finance and Management*, *15*(3), 453–470.

Al-Qahtani, S., & Albakjaji, M. (2023). The role of the legal frameworks in attracting foreign investments: The case of Saudi Arabia. *Asian Journal of Economic and Environmental Journal*. Retrieved from https://ajee-journal.com/the-role-of-the-legal-frameworks-in-attracting-foreign-investments-the-case-of-saudi-arabia

Alshater, M. M., & Othman, A. H. A. (2020). Financial Technology Developments and their Effect on Islamic Finance Education, Journal of King Abdulaziz University: Islamic Economics, King Abdulaziz University. *Islamic Economics Institute.*, *33*(3), 161–187.

Altarturi, B. H. M., Altarturi, H. H. M., & Othman, A. H. A. (2021). Applications of financial technology in Islamic finance: A systematic bibliometric review. In *Artificial Intelligence and Islamic Finance*. Routledge. DOI: 10.4324/9781003171638-10

Altman, E. (1968). Financial ratios, discriminant analysis and the prediction of corporate bankruptcy. Journal of financial, pp 189-209. DOI: 10.1111/j.1540-6261.1968.tb00843.x

Altman, N. S. (1992). An introduction to kernel and nearest-neighbor nonparametric regression. *The American Statistician*, *46*(3), 175–185. DOI: 10.1080/00031305.1992.10475879

Anindyastri, R., Lestari, W. D., Sholahuddin, M. (2022). The Influence of Financial Technology (Fintech) on the Financial Performance of Islamic Banking (Study on Islamic Banking listed on the Indonesia Stock Exchange Period 2016-2020). Benefit: Jurnal Manajemen dan Bisnis, 7(1), 80-92.

Ansori, M. (2019). Perkembangan dan Dampak Financial Technology (Fintech) terhadap Industri Keuangan Syariah di Jawa Tengah. *Wahana Islamika: Jurnal Studi Keislaman*, *5*(1), 32–45.

Arif, M. (1985). Toward the sharia paradigm of Islamic Economics, the beginning of a scientific revolution, *The American of Islamic Social Science*-vol 2.

Asare, N., Alhassan, A. L., Asamoah, M. E., & Ntow-Gyamfi, M. (2017). Intellectual capital and profitability in an emerging insurance market. *Journal of Economic and Administrative Sciences*, *33*(1), 2–9. DOI: 10.1108/JEAS-06-2016-0016

Ashendena, S. K., Bartosik, A., Agapow, P.-M., & Semenova, E. (2021). *Introduction to artificial intelligence and machine learning. The Era of Artificial Intelligence, Machine Learning, and Data Science in the Pharmaceutical Industry*. Elsevier.

Askari, H., & Mirakhor, A. (2017). *Ideal Islamic Economy*. Book.

Askari, H., & Rehman, S. (2010). *How Islamic are Islamic Countries?* (Vol. 10). Global Economy Journal.

Awashreh, R., & Zaabanut, M. (2025). The role of foreign companies operating in Omani freezone in revitalizing small and middle enterprises. In E. AlDhaen, A. Braganza, A. Hamdan, & W. Chen (Eds.), *Business sustainability with artificial intelligence (AI): Challenges and opportunities* (pp. 999–1008). Springer Nature. https://doi.org/DOI: 10.1007/978-3-031-71526-6_89

Awashreh, R. (2025). Leadership strategies for managing change and fostering innovation over private higher education: Academic institutions in Oman. In *Evolving strategies for organizational management and performance evaluation*. IGI Global., DOI: 10.4018/979-8-3373-0149-5.ch018

Awashreh, R., Al-Ghunaimi, H., Saleh, R., & Al-Bahri, M. N. (2024). The impact of leadership roles and strategies on employees' job satisfaction in Oman. *Pakistan Journal of Life and Social Sciences*, *22*(2). Advance online publication. DOI: 10.57239/PJLSS-2024-22.2.00120

Awashreh, R., & Hassiba, A. (2025). Green human resources management for sustainable development in emerging economies. In *Examining green human resources management and nascent entrepreneurship* (pp. 51–78). IGI Global., DOI: 10.4018/979-8-3693-7046-9.ch003

Awashreh, R., & Mohamed, Y. (2024). A comparative study of leadership perceptions in government and education sectors: Insights from the UAE and Oman. *The International Journal of Organizational Diversity*, *24*(2), 109–123. DOI: 10.18848/2328-6261/CGP/v24i02/109-123

Aysan, A. F., Dolgun, M. H., & Turhan, M. İ. (2013). Assessment of the Participation Banks and Their Role in Financial Inclusion in Turkey. *Emerging Markets Finance & Trade*, *49*(5), 99–111. DOI: 10.2753/REE1540-496X4905S506

Baber, H. (2020). FinTech, crowdfunding and customer retention in Islamic banks. *Vision (Basel)*, *24*(3), 260–268. DOI: 10.1177/0972262919869765

Batir, T. E., Volkman, D. A., & Gungor, B. (2017). Determinants of Bank Efficiency in Turkey: Participation Banks Versus Conventional Banks. *Borsa Istanbul Review*, *17*(2), 86–96. DOI: 10.1016/j.bir.2017.02.003

Baykuş, O., & Bektaş, S. (2021, August). Katılım Bankalarının Kâr Paylaşım Oranlarını Belirleyen Etmenler Üzerine Ampirik Bir İnceleme: Türkiye Katılım Bankaları Örneği. *The Journal of Accounting and Finance*, (Special Issue), 397–422.

Beik, I. S., & Arsyianti, L. D. (2021). Digital Technology and Its Impact on Islamic Social Finance Literacy, Springer Books, in: Mohd Ma'Sum Billah (ed.), Islamic FinTech, edition 1, pp. 429-445, Springer.

Benbba, R., Barhdadi, M., Ficarella, A., Manente, G., Romano, M. P., El Hachemi, N., Barhdadi, A., Al-Salaymeh, A., & Outzourhit, A. (2024). Solar energy resource and power generation in Morocco: Current situation, potential, and future perspective. *Resources*, *13*(10), 140. DOI: 10.3390/resources13100140

Benhamed, A., & Gassouma, M. S. (2023). Preventing Oil Shock Inflation: Sustainable Development Mechanisms vs. Islamic Mechanisms. *Sustainability (Basel)*, *15*(12), 2. DOI: 10.3390/su15129837

Benzaken, D., Adam, J. P., Virdin, J., & Voyer, M. (2024). From concept to practice: Financing sustainable blue economy in Small Island Developing States, lessons learnt from the Seychelles experience. *Marine Policy*, *163*, 106072. Advance online publication. DOI: 10.1016/j.marpol.2024.106072

Berthod, O. (2018). Institutional theory of organizations. In Farazmand, A. (Ed.), *Global Encyclopedia of Public Administration, Public Policy, and Governance* (pp. 1–5). Springer., DOI: 10.1007/978-3-319-20928-9_63

Biancone, P., Uluyol, B., Petricean, D., & Chmet, F. (2020). The Bibliometric Analysis of Islamic Banking and Finance. *Journal of Islamic Accounting and Business Research*, *11*(10), 2069–2086. DOI: 10.1108/JIABR-08-2020-0235

Boscia, V., Stefanelli, V., Coluccia, B., & De Leo, F. (2019). The role of finance in environmental protection: A report on regulators' perspective. *Risk Governance and Control: Financial Markets & Institutions*, *9*(4), 30–40. DOI: 10.22495/rgcv9i4p3

BRSA. (2025). Monthly Banking Sector Data (Basic Analysis), https://www.bddk.org.tr/BultenAylik/en

Bruce, C. (1994). Supervising LRPs. UK. In Zuber-Skerritt, O., & Ryan, Y. (Eds.), *Quality in postgraduate education*. Kogan Page.

BSI. (2015). *Architectural framework for the Internet of Things, for Smart Cities*. BSI.

Burton, B., & Burke, B. (2012). *EA in the Arab Gulf States: Increased Focus on Delivering Strategic Value. Published: 3 August 2012*. Gartner Inc.

Butryna, B., Chomiak-Orsa, I., Hauke, K., Pondel, M., & Siennicka, A. (2021). Application of Machine Learning in medical data analysis illustrated with an example of association rules. *Procedia Computer Science*, *192*, 3134–3143. DOI: 10.1016/j.procs.2021.09.086

Carè, R., & De Lisa, R. (2019). Social impact bonds for a sustainable welfare state: The role of enabling factors. *Sustainability (Basel)*, *11*(10), 2884. DOI: 10.3390/su11102884

Cengiz, S., & Özkan, T. (2023). The Place of FinTech Applications in Islamic Finance: A Conceptual Evaluation. *Journal of Ilahiyat Researches*, *60*(1), 1–14.

Çevik, S., & Charap, J. (2011). *The Behavior of Conventional and Islamic Bank Deposit Returns in Malaysia and Turkey*. IMF Working Paper, WP/11/156.

Chammas, G. (2006). Islamic Finance Industry In Lebanon: Horizons, Enhancements And Projections. A thesis. Ecole Supérieure des Affaires (ESA)-Lebanon.

Chapra, M. U. (1982). Money and Banking in an Islamic Economy. In Ariff, M. (Ed.), *Monetary and Fiscal Economics of Islam*. International Centre for Research in Islamic Economics.

Charnes, A., Cooper, W. W., & Rhodes, E. (1978). Measuring the efficiency of decision making units. *European Journal of Operational Research*, *2*(6), 429–444. DOI: 10.1016/0377-2217(78)90138-8

Chedid, E. (2018). Regulatory updates on Islamic banking in Lebanon. Dentons. Lebanon. https://www.dentons.com/en/insights/alerts/2018/august/15/regulatory-updates-on-islamic-banking-in-lebanon

Chen, Y., Zhang, X., & Wang, Q. (2021). The Old Boys Club in New Zealand Listed Companies. *Journal of Risk and Financial Management, 14*(8), 345. DOI: 10.3390/jrfm14080342

Chong, F. H. L. (2021). Enhancing trust through digital Islamic finance and blockchain technology. *Qualitative Research in Financial Markets, 13*(3), 328341. DOI: 10.1108/QRFM-05-2020-0076

Clarke, Th., & Tigue, J. (1975). *Dirty money: Swiss banks, the Mafia, money laundering, and white-collar crime*. Simon and Schuster.

Clark, R., Reed, J., & Sunderland, T. (2018). Bridging funding gaps for climate and sustainable development: Pitfalls, progress, and potential of private finance. *Land Use Policy, 71*, 335–346. DOI: 10.1016/j.landusepol.2017.12.013

Coskun, Ş., Turanlı, M., & Yılmaz, K. (2024). The Relationship of Funds Collected and Disbursed in Participation Banks with Sector Shares and the Effect of Macroeconomic Indicators. *Journal of Islamic Research, 35*(2), 240–253.

Courrier International. (2024). Crise. Le "Madoff libanais" Riad Salamé enfin derrière les barreaux: un "tour de passe-passe"? Courrier International. https://www.courrierinternational.com/article/crise-le-madoff-libanais-riad-salame-enfin-derriere-les-barreaux-un-tour-de-passe-passe_221826

Cumhurbaşkanlığı Finans Ofisi. (2023). *Fintek sözlüğü*. cbfo.gov.tr/

Daryanto, W. M., Akbar, F., & Perdana, F. A. (2020). Financial performance analysis in the banking sector: Before and after financial technology regulation in Indonesia (case study of Buku-iv in Indonesia for period 2013-2019). *International Journal of Business. Economics and Law, 21*(2), 1–9.

De Anca, C. (2019). Fintech in Islamic Finance: From collaborative finance to community-based finance. In *Fintech in Islamic Finance* (pp. 47–61). Routledge. DOI: 10.4324/9781351025584-4

Delle Foglie, A., Panetta, I. C., Boukrami, E., & Vento, G. (2021). The impact of the Blockchain technology on the global Sukuk industry: Smart contracts and asset tokenisation. *Technology Analysis and Strategic Management, 1*(1), 1–15.

Demirdöğen, Y. (2023). Blockchain ve İslami Finans: Türkiye'de Uygulamalar ve Gelecek Perspektifleri. *Türkiye İslami Fintech Araştırmaları Dergisi*, 30-48.

Demirguc-Kunt, A., Klapper, L., Singer, D., Ansar, S., & Hess, J. (2018). The Global Findex Database 2017: Measuring Financial Inclusion and the Fintech Revolution. In *Washington, DC: World Bank eBooks*. DOI: 10.1596/978-1-4648-1259-0

DinarStandart. (2018). "Islamic Fintech Report 2018: Current Lanscape and Path Forward." Dubai Islamic Economy Development Centre 1-38.

Dirie, K. A., Alam, M. M., & Maamor, S. (2023). Islamic social finance for achieving sustainable development goals: A systematic literature review and future research agenda. *International Journal of Ethics and Systems*, *40*(1), 676–698. Advance online publication. DOI: 10.1108/IJOES-12-2022-0317

Dushkova, D., & Kuhlicke, C. (2024). Making co-creation operational: A RECONECT seven-steps-pathway and practical guide for co-creating nature-based solutions. *MethodsX*, *12*, 102495. DOI: 10.1016/j.mex.2023.102495 PMID: 38170128

Dzinamarira, T., Bopoto, E., Mutiro, T., Chitungo, I., & Dzinamarira, T. (2024). Overcoming challenges and thriving as a new NGO – a case for institutional capacity building. *Development in Practice*, *34*(8), 949–953. DOI: 10.1080/09614524.2024.2358325

Eke, V., & Sevinç, H. (2021). Türkiye'deki Katılım Bankalarının Etkinlik Analizi: Özel ve Kamu Bankalarının Karşılaştırılması. *Iğdır Üniversitesi Sosyal Bilimler Dergisi*, (28), 434–451.

Elamin, M. O. (2023). Advancing ethical and sustainable economy: Islamic finance solutions for environmental, social, & economic challenges in the digital age. *International Journal of Membrane Science and Technology*, *10*(5), 408–429. DOI: 10.15379/ijmst.v10i5.2515

El-Gamal, M. A. (2006). Islamic Finance: Law, Economics and Practice. *Cambridge University Press*, 221. DOI: 10.4197/islec.21-2.5

El-Gamal, M. (2020). *Islamic Finance: Law, Economics, and Practice*. Cambridge University Press.

Elmas, B., & Yetim, A. (2021). Katılım Bankalarının Finansal Performanslarının TOPSIS Yöntemi İle Uluslararası Boyutta Değerlendirilmesi. *Uluslararası İslam Ekonomisi ve Finansı Araştırmaları Dergisi*, *7*(3), 230–263.

Emeç, Ö. (2020). *Yeni Dünya ve Yeni Finans: Ortaklık Temelli Finansman ve Katılım Bankaları*. Albaraka Yayınları.

Eweje, G., Sajjad, A., Nath, S. D., & Kobayashi, K. (2020). Multi-stakeholder partnerships: A catalyst to achieve sustainable development goals. *Marketing Intelligence & Planning*. Advance online publication. DOI: 10.1108/MIP-04-2020-0135

Falcone, P. M. (2020). Environmental regulation and green investments: The role of green finance. *International Journal of Green Economics*, *14*(2), 159–173. DOI: 10.1504/IJGE.2020.109735

Farhoomand, A., & Lentini, D. (2008). e-Business Transformation in the Banking Industry: The Case of Citibank. Asia Case Research Centre. The University of Hong Kong. Retrieved from http://www.acrc.hku.hk/case/case_showdetails.asp?ct=newly&c=944&cp=1949&pt=1

Faster Capital. (2024). *İslami Fintech: Dijital İslami Bankacılık Hizmetlerinde Yenilikler*. https://fastercapital.com/: https://fastercapital.com/content/Islamic-Fintech--Innovations-in-Digital-Islamic-Banking Services.html#:~:text=The%20utilization%20of%20blockchain%20technology,issuance%2C%20and%20Takaful%20

Finterra. (2024, January 13). Islamic FinTech – An Evolution, or a Revolution! - The Finterra Publication - Medium. *Medium*. https://medium.com/finterra/islamic-fintech-an-evolution-or-a-revolution-bf9a80b27e36

Fisk, R. (2011). Robert Fisk: Phoenicians footprints all over Beirut. https://www.independent.co.uk/voices/commentators/fisk/robert-fisk-phoenician-footprints-all-over-beirut-6271510.html. Independent.

Fu, C., Lu, L., & Pirabi, M. (2023). Advancing green finance: A review of sustainable development. *DESD*, *1*(1), 20. DOI: 10.1007/s44265-023-00020-3

G20 Sustainable Finance Working Group. (2023). *Sustainable finance report: Volume I*. https://g20sfwg.org/wp-content/uploads/2023/10/Volume-I-G20-India-Final-VF.pdf

Gabrys, E., & Field, S. Gartner, (2014). CIO Agenda: A Gulf Cooperation Council Perspective. Published: 20 March 2014. Gartner Inc. USA.

Gao, Y., Li, S., & Zhou, H. (2023). *Journal of Financial Services Research, 61*(1), 89–112.

Gassouma, M. S. (2022). Modeling and Conceptualization of an Islamic Financial System, 5th International Scientific Conference Islam and Contemporary Issues, Turkey – Antalya, 2022, Euro-Arab Organization for Water and Desert & University of Zitouna.

Gassouma, M. S. (2023.a). Empirical Modeling of Sharia Compliance in Islamic Financial Transactions: A Case Study on the Alignment of Islamic and Conventional Finance with Sharia Standards, 6th International Scientific Conference Islam and Contemporary Issues, Turkey – Antalya, 2023, Euro-Arab Organization for Water and Desert & University of Zitouna.

Gassouma, M. S. (2023a). Empirical Modeling of Sharia Compliance in Islamic Financial Transactions: A Case Study on the Alignment of Islamic and Conventional Finance with Sharia Standards, 6th International Scientific Conference Islam and Contemporary Issues, Turkey – Antalya. Euro-Arab Organization for Water and Desert & University of Zitouna.

Gassouma, M.S and Rajhi, M.T. (2007). Evaluation de Data Mining pour la classification des entreprises industrielles Tunisiennes dans le cadre du credit scoring. Revue de l'association francophone de management électronique.

Gassouma, M.S., Benahmed, A., Montasser, G. (2022). Investigating similarities between Islamic and conventional banks in GCC countries: a dynamic time warping approach. *International Journal of Islamic and Middle Eastern Finance and Management.*

Gassouma, M.S., Ghroubi, M. (2021). Discriminating between Islamic and Conventional banks in term of cost efficiency with combination of credit risk and interest rate margin in the GCC countries: Does Arab revolution matter? *ACRN Oxford Journal of Finance and Risk Perspectives, 10.*

Gassouma, M. S. (2023). Applications of an Alternative Islamic Profit Rate to Interest Rates: Examining Its Impact on Investment and Inflation, *2nd International Conference on Islamic Finance*, Tunisia – Hammamet, University of Zitouna.

Gassouma, M. S., Benahmed, A., & Montasser, G. (2022). (in press). a). Investigating similarities between Islamic and conventional banks in GCC countries: A dynamic time warping approach. *International Journal of Islamic and Middle Eastern Finance and Management.*

Gateway, S. (2019). *Global Islamic Fintech Report 2019 | Salaam Gateway - Global Islamic Economy Gateway.* Salaam Gateway - Global Islamic Economy Gateway. https://salaamgateway.com/specialcoverage/islamic-fintech-2019

Gateway, S. (2024). *Global Islamic Fintech Report 2023/24 | Salaam Gateway - Global Islamic Economy Gateway.* Salaam Gateway - Global Islamic Economy Gateway. https://salaamgateway.com/specialcoverage/islamic-fintech-2023

Gateway, S. (2024). *Top 30 Digital Islamic Economy Startups 2024.* Salaam Gateway.

Geranmayeh, E. (2018). *Regional Geopolitical Rivalries in the Middle East: Implications for Europe. IAI-FEPS. Istituto Affari Internazionali (IAI) and Foundation for European Progressive Studies.* FEPS.

Gomart, Th. (2016). *The Return Of Geopolitical Risk-Russia, China and the United States. Institut français des relations internationales.* Ifri.

Gorelick, J., Cara, E., & Kavoo, G. (2024). The fallacy of green municipal bonds in developing countries. *WORLD (Oakland, Calif.)*, *5*(4), 929–951. DOI: 10.3390/world5040047

Goud, B., Uddin, T. A., & Fianto, B. A. (2021). Islamic Fintech and ESG goals: Key considerations for fulfilling maqasid principles. In *Islamic Fintech* (pp. 16–35). Routledge.

Grand View Research. (2023). Cryptocurrency Market Size, Share & Growth Report, 2030. Retrieved May 14, 2024, from https://www.grandviewresearch.com/industry-analysis/cryptocurrency-market-report

Greggwirth. (2022). *The ESG potential of Islamic finance - Thomson Reuters Institute*. Thomson Reuters Institute. https://www.thomsonreuters.com/en-us/posts/news-and-media/islamic-finance-esg/

Griffiths, S., & Sovacool, B. K. (2020). Rethinking the future low-carbon city: Carbon neutrality, green design, and sustainability tensions in the making of Masdar City. *Energy Research & Social Science*, *62*, 101368. DOI: 10.1016/j.erss.2019.101368

GTR. (2024, April 22). How sustainability is shaping trade finance in the Middle East. *Global Trade Review*. https://www.gtreview.com/magazine/the-esg-trade-issue-2024/how-sustainability-is-shaping-trade-finance-in-the-middle-east/

Güler, M., Kabakçı, A., Koç, Ö., Eraslan, E., Derin, K. H., Güler, M., Ünlü, R., Türkan, Y. S., & Namlı, E. (2024). Forecasting of the Unemployment Rate in Turkey: Comparison of the Machine Learning Models. *Sustainability (Basel)*, *16*(15), 6509. DOI: 10.3390/su16156509

Güler, M., & Namlı, E. (2024). Brain Tumor Detection with Deep Learning Methods' Classifier Optimization Using Medical Images. *Applied Sciences (Basel, Switzerland)*, *14*(2), 642. DOI: 10.3390/app14020642

Gunasekare, U. (2015). Mixed Research Method as the Third Research Paradigm: A Literature Review. Volume 4 Issue 8, August 2015. University of Kelaniya. IJSR.

Gürçay, H. R., & Dağıdır, C. (2022). COVID-19 Sürecinde Katılım Bankaları ile Özel Mevduat Bankalarının Performans Değerlendirmesi: Türkiye Örneği. *Uluslararası Finansal Ekonomi ve Bankacılık Uygulamaları Dergisi*, *3*(1), 1–25.

Hacak, H., & Gürbüz, Y. (2019). İslami Finansta Sigorta ve Katılım Sigortası (Tekâfül)", Yaşayan ve gelişen katılım bankacılığı. *TKBB (Türkiye Katılım Bankaları Birliği) Yayınları*, 300-317.

Hamadou, I., & Suleman, U. (2024). FinTech and Islamic Finance: Opportunities and Challenges, Smolo, E. and Raheem, M.M. (Ed.) The Future of Islamic Finance, Emerald Publishing Limited, Leeds, pp. 175-188.

Hamadou, I., Yumna, A., Hamadou, H., & Jallow, M. (2024). Unleashing the power of artificial intelligence in Islamic banking: A case study of Bank Syariah Indonesia (BSI). MF-Journal. https://mf-journal.com/article/view/116

Hamarat, Ç. (2024). The Efficiency of Participation and Conventional Banking in Turkiye: A Stochastic Frontier Approach. *Toplum Ekonomi ve Yönetim Dergisi*, *5*(1), 56–79. DOI: 10.58702/teyd.1341253

Hand, D. J., Till, R. J., & Patil, S. (2001). A simple generalization of the area under the ROC curve for multiple class classification problems. *Machine Learning*, *45*(2), 171–186. DOI: 10.1023/A:1010920819831

Hanif, M. (2011). Differences and Similarities in Islamic and Conventional Banking. *International Journal of Business and Social Science*, *2*(2), 166–175.

Han, J., & Kamber, M. (2006). *Data Mining: Concepts and Techniques*. Morgan Kaufmann.

Han, J., Pei, J., & Yin, Y. (2000). Mining frequent patterns without candidate generation. In *Proceedings of the 2000 ACM SIGMOD international conference on Management of data* (pp. 1-12).

Haqqi, A. R. A. (2020). Strengthening Islamic Finance in South-East Asia Through Innovation of Islamic FinTech in Brunei Darussalam. In *Economics, Business, and Islamic Finance in ASEAN Economics Community* (pp. 202-226). IGI Global.

Hasan, M., & Dridi, J. (2010). *The Effects of the Global Crisis on Islamic and Conventional Banks: A Comparative Study*. IMF Working Paper, WP/10/201.

Hasan, M. B., Rashid, M. M., Shafiullah, M., & Sarker, T. (2022). How resilient are Islamic financial markets during the COVID-19 pandemic? *Pacific-Basin Finance Journal*, *74*, 101817. DOI: 10.1016/j.pacfin.2022.101817

Hasan, R., Hassan, M. K., & Aliyu, S. (2020). Fintech and Islamic finance: Literature review and research agenda. [IJIEF]. *International Journal of Islamic Economics and Finance*, *3*(1), 75–94. DOI: 10.18196/ijief.2122

Hasnan, B. (2019). Financial inclusion and FinTech: A comparative study of countries following Islamic finance and conventional finance. *Qualitative Research in Financial Markets, 12*(1), 24–42. DOI: 10.1108/QRFM-12-2018-0131

Hassan, K., Rabbani, M. R., & Mahmood, A. M. A. (2020). Challenges for the Islamic Finance and banking in post COVID era and the role of Fintech. *Journal of Economic Cooperation and Development, 41*, 93–116.

Hassan, K., Rabbani, M. R., Rashid, M., & Trinugroho, I. (2022). Islamic Fintech, Blockchain and Crowdfunding: Current Landscape and Path Forward. In *FinTech in Islamic Financial Institutions: Scope, Challenges, and Implications in Islamic Finance* (pp. 307–340). Springer International Publishing. DOI: 10.1007/978-3-031-14941-2_15

Hassan, M. K., Aliyu, S., Huda, M., & Rashid, M. (2019). A Survey on Islamic Finance and Acconting Standarts. *Borsa İstanbul Review, 19*(1), 1–13. DOI: 10.1016/j.bir.2019.07.006

Hendratmi, A., Ryandono, M. N. H., & Sukmaningrum, P. S. (2020). Developing Islamic crowdfunding website platform for startup companies in Indonesia. *Journal of Islamic Marketing, 11*(5), 1041–1053. DOI: 10.1108/JIMA-02-2019-0022

Hicks, J. R. (1937). Mr. Keynes and the 'Classics': A Suggested Interpretation. *Econometrica, 5*(2), 147–159. DOI: 10.2307/1907242

Hidajat, T. (2020). Financial Technology in Islamic View. *Perisai: Islamic Banking and Finance Journal, 4*(2), 102–112. DOI: 10.21070/perisai.v4i2.465

Hodge, G., & Greve, C. (2007). Public–private partnerships: An international performance review. *Public Administration Review, 67*(3), 545–558. DOI: 10.1111/j.1540-6210.2007.00736.x

Hud Saleh Huddin, M., Lee, M. ve Mansor, M. S. (2022). Islamic fintech nascent and on the rise. IADI Fintech Brief, 1-18.

Hudaefi, F. A. (2020). How does Islamic fintech promote the SDGs? Qualitative evidence from Indonesia. *Qual. Res. Financial Markets, 12*(4), 353–366. DOI: 10.1108/QRFM-05-2019-0058

Hudaefi, F. A., Hassan, M. K., & Abduh, M. (2023). Exploring the development of Islamic fintech ecosystem in Indonesial. a text analytics. *Qualitative Research in Financial Markets, 15*(3), 514–533. DOI: 10.1108/QRFM-04-2022-0058

Hussain, M., Shahmoradi, A., & Turk, R. (2015). *An Overview of Islamic Finance.* IMF Working Paper, WP/15/120.

Hussain, M., & Imran, M. (2020). *Research in International Business and Finance, 54*, 101263.

Hussain, M., Nadeem, M. W., Iqbal, S., Mehrban, S., Fatima, S. N., Hakeem, O., & Mustafa, G. (2019). Security and Privacy in FinTech: A Policy Enforcement Framework. In Rafay, A. (Ed.), *FinTech as a Disruptive Technology for Financial Institutions* (pp. 81–97). IGI Global. DOI: 10.4018/978-1-5225-7805-5.ch005

Iao-Jörgensen, J. (2024). Networking in action: Taking collaborative capacity development seriously for disaster risk management. *Progress in Disaster Science, 21*, 100311. DOI: 10.1016/j.pdisas.2024.100311

Ibn khaldoun (1377). El Mukaddimah. Book.

Ibrahim, A. S., Zaki, A., & Al-R., A. (2023). Applied Ijtihad of Sharia Rulings: Contemporary Financial Transactions as a Model. *IUG Journal of Sharia & Law Studies, 31*(1).

Ibrahim, M., & Mohamed, N. A. E. M. (2016). Towards sustainable management of solid waste in Egypt. *Procedia Environmental Sciences, 34*, 336–347. DOI: 10.1016/j.proenv.2016.04.030

IFN. (2024). IFNFintech on the pulse of islamic fintech, ifnfintech.com/landscape/ https://ifnfintech.com/landscape/

IFSB. (2024). *Islamic Financial Services Industry Stabiliity Report 2024.* https://www.ifsb.org/wp-content/uploads/2024/09/IFSB-Stability-Report-2024-8.pdf

Ikevuje, A. H. U., Anaba, D. C., & Iheanyichukwu, U. T.Augusta Heavens IkevujeDavid Chinalu AnabaUche Thankgod Iheanyichukwu. (2024). Exploring sustainable finance mechanisms for green energy transition: A comprehensive review and analysis. *Finance & Accounting Research Journal, 6*(7), 1224–1247. DOI: 10.51594/farj.v6i7.1314

IMF. (2017). *Ensuring Financial Stability in Countries with Islamic Banking-Country Case Studies.* IMF Country Report No. 17/145.

Iqbal, M., & Molyneux, P. (2021). *Islamic Finance: Principles and Practice*. Routledge.

Jonkers, H., Band, I., & Quartel, D. (2012a). *ArchiSurance Case Study*. The Open Group.

Joseph, Ch. (2014). Types of eCommerce Business Models. https://smallbusiness.chron.com/types-ecommerce-business-models-2447.html. Demand Media.

Jung, H. (2020). Development finance, blended finance, and insurance. *International Trade. Politics and Development, 4*(1), 47–60. DOI: 10.1108/ITPD-12-2019-0011

Kagan, J. (2020). Financial Technology – FinTech. Investopedia. https://www.investopedia.com/terms/f/FinTech.asp

Kamalov, F., Santandreu Calonge, D., & Gurrib, I. (2023). New era of artificial intelligence in education: Towards a sustainable multifaceted revolution. *Sustainability (Basel), 15*(16), 12451. DOI: 10.3390/su151612451

Kannaiah, D., Masvood, Y., & Choudary, Y. L. (2017). Growth of Islamic banking in India: Discriminant analysis approach. *Banks Bank Syst., 12*(4), 175–188. DOI: 10.21511/bbs.12(4-1).2017.06

Karapinar, A., & Dogan, İ. C. (2015). An Analysis on the Performance of the Participation Banks in Turkey. *Accounting and Finance Research, 4*(2), 24–33. DOI: 10.5430/afr.v4n2p24

Karim, R. A., & Archer, S. (2013). *Islamic Finance: The New Regulatory Challenge.* Wiley Finance. https://www.wiley.com/en-ae/Islamic+Finance%3A+The+New+Regulatory+Challenge%2C+2nd+Edition-p-9781118628973

Karim, A. K. M. Rezaul, Hasan, M. (2019). Islamic Finance and Fintech: A Critical Review. *Journal of Islamic Accounting and Business Research*, *10*, 296–311.

Karim, S., Naeem, M. A., & Abaji, E. E. (2022). Is Islamic FinTech coherent with Islamic banking? A stakeholder's perspective during COVID-19. *Heliyon*, *8*(9), 1–6. DOI: 10.1016/j.heliyon.2022.e10485 PMID: 36110236

Kartal, F. (2012). Interest-Free Banking in the World and a Financial Analysis of the Turkey Experience. *International Research Journal of Finance and Economics*, *93*, 183–201.

Kassir, K. (2010). *Beirut*. University of California Press.

Keraine, R. (2019). The Fundamentals of Islamic Finance. INVIVOO. https://blog.invivoo.com/the-fundamentals-of-islamic-finance/

Keynes, J. M. (1936). *The General Theory of Employment, Interest and Money*. Book.

Khan, M. I., Bicer, Y., Asif, M., Al-Ansari, T. A., Khan, M., Kurniawan, T. A., & Al-Ghamdi, S. G. (2024). The GCC's path to a sustainable future: Navigating the barriers to the adoption of energy efficiency measures in the built environment. *Energy Conversion and Management*, *X*, 100636.

Khan, M. S. (1986). Islamic Interest-Free Banking: A Theoretical Analysis. *Staff Papers - International Monetary Fund. International Monetary Fund*, *33*(1), 1–27. DOI: 10.2307/3866920

Khan, M. S., & Mirakhor, M. (1989). The financial system of monetary policy in an Islamic Economy. *Islamic Economic, Vol*, *1*(1), 39–57. DOI: 10.4197/islec.1-1.2

Khan, M. S., Rabbani, M. R., Hawaldar, I. T., & Bashar, A. (2022). Determinants of behavioural intentions to use Islamic financial technology: An empirical assessment. *Risks*, *10*(6), 1–19. DOI: 10.3390/risks10060114

Khiari, W., Ben Flah, I., Lajmi, A., & Bouhleli, F. (2024). The stock market reaction to green bond issuance: A study based on a multidimensional scaling approach. *Journal of Risk and Financial Management*, *17*(9), 408. Advance online publication. DOI: 10.3390/jrfm17090408

Kılıç, G., & Türkan, Y. (2023). The Emergence of Islamic Fintech and Its Applications. *International Journal of Islamic Economics and Finance Studies*, *9*(2), 212–236. DOI: 10.54427/ijisef.1328087

Kismawadi, R., Aditchere, J., & Libeesh, P. (2024). Integration of Artificial Intelligence Technology in Islamic Financial Risk Management for Sustainable Development. Applications of Block Chain technology and Artificial Intelligence. pp 53–71. https://link.springer.com/chapter/10.1007/978-3-031-47324-1_4

Kismawadi, E., Irfan, M., Abdul, S., & Shah, R. (2023). *Revolutionizing Islamic Finance: Artificial Intelligence's Role in the Future of Industry. Book: The Impact of AI Innovation on Financial Sectors in the Era of Industry 5.0 (/book/impact-innovation-financial-sectors-era/321132).* IGI-Global., DOI: 10.4018/979-8-3693-0082-4.ch011

Kok, S. K., Akwei, C., Giorgioni, G., & Farquhar, S. (2022). On the regulation of the intersection between religion and the provision of financial services: Conversations with market actors within the global Islamic financial services sector. *Research in International Business and Finance*, *59*, 101552. DOI: 10.1016/j.ribaf.2021.101552

Kolbe, Th. (2015). Smart Models for Smart Cities - Modeling of Dynamics, Sensors, Urban Indicators, and Planning Actions. 29th of October 2015 Joint International Geoinformation Conference JIGC 2015, Kuala Lumpur.

Kothari, S. P., Leone, A. J., & Wasley, C. E. (2005). Performance matched discretionary accrual measures. *Journal of Accounting and Economics*, *39*(1), 163–197. DOI: 10.1016/j.jacceco.2004.11.002

Kowall, J., & Fletcher, C. (2013). *Modernize Your Monitoring Strategy by Combining Unified Monitoring and Log Analytics Tools*. Gartner Inc.

KPMG. (2020). Top 10 FinTech predictions for 2020. KPMG. https://home.kpmg/xx/en/home/campaigns/2020/02/pulse-of-FinTech-h2-19-top-10-predictions-for-2020.html

Kshatri, S. S., Singh, D., Goswami, T., & Sinha, G. R. (2023). *Introduction to statistical modeling in machine learning: A case study. Statistical Modeling in Machine Learning*. Elsevier. DOI: 10.1016/B978-0-323-91776-6.00007-5

Kuanova, L. A., Sagiyeva, R., & Shah Shirazi, N. (2021). Islamic social finance: A literature review and future research directions. *Journal of Islamic Accounting and Business Research, 12*(5), 707728. DOI: 10.1108/JIABR-11-2020-0356

Kühlert, M., Klingen, J., Gröne, K., Hennes, L., Terrapon-Pfaff, J. C., & Jamea, E. M. (2024, February). Pathways towards a green economy in Egypt: Final report. Wuppertal Institute for Climate, Environment and Energy. https://epub.wupperinst.org/frontdoor/deliver/index/docId/8534/file/8534_Green_Economy_Egypt.pdf

Lacasse, R.-M., Lambert, B., & Khan, N. (2018). Islamic Banking - Towards a Blockchain Monitoring Process. *5th International Conference on Entrepreneurial Finance* (s. 33-46). Morocco: Journal of Business and Economics.

Lahmiri, S., & Bekiros, S. (2019). Decomposing the persistence structure of Islamic and green crypto-currencies with nonlinear stepwise filtering. Chaos, Solit. *Chaos, Solitons, and Fractals, 127*, 334–341. DOI: 10.1016/j.chaos.2019.07.012

Lazar, I., Motogna, S., & Parv, B. (2010). Behaviour-Driven Development of Foundational UML Components. Department of Computer Science. Babes-Bolyai University. Cluj-Napoca, Romania. DOI: 10.1016/j.entcs.2010.07.007

Le Monde. (2019). Le Prix Nobel d'économie Angus Deaton: Quand l'Etat produit une élite prédatrice [Nobel Lauréate in Economics Angus Deaton: "When the state produces a predatory elite]. Le Monde. Retrieved from https://www.lemonde.fr/idees/article/2019/12/27/angus-deaton -quand-l-etat-produit-une-elite-predatrice_6024205_3232.html

Lea, R. (2017). Smart City Standards: An overview.

Legowo, M. B., Subanidja, S., & Sorongan, F. A. (2021). Fintech and bank: Past, present, and future. *Jurnal Teknik Komputer AMIK BSI, 7*(1), 94–99. DOI: 10.31294/jtk.v7i1.9726

Liu, Y., & Huang, R. (2021). *International Journal of Productivity and Performance Management, 70*(1), 124–142.

Makarchenko, M., Nerkararian, S., & Shmeleva, S. (2016). How Traditional Banks Should Work in Smart City. Communications in Computer and Information Science. DOI: . Conference: International Conference on Digital Transformation and Global Society.DOI: 10.1007/978-3-319-49700-6_13

Malhotra, E. (2005). Integrating knowledge management technologies in organizational business processes: getting real time enterprises to deliver real business performance. Emerald Group Publishing Limited, Journal of Knowledge Management. USA.

Marin, P. (2009). Public-private partnerships for urban water utilities. *World Bank.* https://ppp.worldbank.org/public-private partnership/sites/ppp.worldbank.org/files/documents/FINAL-PPPsforUrbanWaterUtilities-PhMarin.pdf

Markides, C. (2011, March). Crossing the Chasm: How to Convert Relevant Research Into Managerially Useful Research. [London, UK.]. *The Journal of Applied Behavioral Science, 47*(1), 121–134. DOI: 10.1177/0021886310388162

Martínez-Peláez, R., Ochoa-Brust, A., Rivera, S., Félix, V. G., Ostos, R., Brito, H., Félix, R. A., & Mena, L. J. (2023). Role of digital transformation for achieving sustainability: Mediated role of stakeholders, key capabilities, and technology. *Sustainability (Basel), 15*(14), 11221. DOI: 10.3390/su151411221

Mashhour, A. (1991). *Investment in Islamic Economics, Madbouly Library.* Arabic.

MENA-OECD. (2018). *Background Note: Country case studies: Building economic resilience in Lebanon and Libya. Resilience in fragile situations. Mena-oecd economic/resilience task force.* Islamic Development Bank.

Mertzanis, C., & Tebourbi, I. (2024). Geopolitical risk and global green bond market growth. *European Financial Management.* Advance online publication. DOI: 10.1111/eufm.12484

Miori, V., & Russo, D. (2014). Domotic Evolution towards the IoT. IEEE. *28th International Conference on Advanced Information Networking and Applications Workshops.* DOI: . Victoria, BC, Canada.DOI: 10.1109/WAINA.2014.128

Mohammad, S., Yasin, M. M., & Ahmad, I. (2023). *Asia Pacific Journal of Management*, *40*(1), 211–229.

Morrison, M. (2016). Critical Success Factors – Analysis made easy, a step by step guide. rapidBI.

Nabi, G., Islam, A., Bakar, R., & Nabi, R. (2017). Islamic microfinance as a tool of financial inclusion in Bangladesh. *J. Islam. Econ. Banking Finance*, *13*(1), 24–51. DOI: 10.12816/0051154

Nakamoto, S. (2008). Bitcoin: A Peer-to-Peer Electronic Cash System. www.bitcoin.org. BITCOIN.

Nanopoulos, A., Papadopoulos, A. N., & Manolopoulos, Y. (2007). Mining association rules in very large clustered domains. *Information Systems*, *32*(6), 649–669. DOI: 10.1016/j.is.2006.04.002

Narayan, P. K., & Phan, D. H. B. (2019). A survey of Islamic banking and finance literature: Issues, challenges and future directions. *Pacific-Basin Finance Journal*, *53*, 484–496. DOI: 10.1016/j.pacfin.2017.06.006

Nguyen, H. T., & Tran, P. Q. (2021). *Journal of Banking & Finance*, *124*, 106038.

Noronha, M. (2020, April 27). *Islamic fintech: Reaching the next generation of Muslims.* Economist İmpact: https://impact.economist.com/perspectives/financial-services/islamic-fintech-reaching-next-generation-muslims

Nurdin, N., & Khaeruddin, Y. (2020). Knowledge management lifecycle in Islamic bank: The case of syariah banks in Indonesia. *International Journal of Knowledge Management Studies*, *11*(1), 59–80. DOI: 10.1504/IJKMS.2020.105073

Nurhadi, N. (2019). The importance of maqashid sharia as a theory in Islamic economic business operations. [IJIBEC]. *International Journal of Islamic Business and Economics*, *3*(2), 130–145. DOI: 10.28918/ijibec.v3i2.1635

OASIS. (2014). *ISO/IEC and OASIS Collaborate on E-Business Standards-Standards Groups Increase Cross-Participation to Enhance Interoperability*. The OASIS Group.

Oberauer, N. (2018). Money in classical Islam: Legal theory and economic practice. *Islam Law Soc.*, *25*(4), 427–466. DOI: 10.1163/15685195-00254A03

OECD. (2020). *Blended finance in the least developed countries 2020: Supporting a resilient COVID-19 recovery*. Organisation for Economic Co-operation and Development.

Oladapo, I. A., Hamoudah, M. M., Alam, M. M., Olaopa, O. R., & Muda, R. (2022). Customers' perceptions of FinTech adaptability in the Islamic banking sector: Comparative study on Malaysia and Saudi Arabia. *Journal of Modelling in Management*, *17*(4), 1241–1261. DOI: 10.1108/JM2-10-2020-0256

OLJ. (2023). Des centaines de millions de dollars que Riad Salamé est accusé d'avoir détournés auraient atterri en Suisse. https://www.lorientlejour.com/article/1329590/les-fonds-detournees-de-la-banque-centrale-du-liban-ont-ete-transferes-en-suisse-media.html

OLJ. (2024). Crimes Financiers Au Liban-Mandat d'arrêt contre Riad Salamé, qui reste en détention à l'issue de son interrogatoire/La cheffe du contentieux de l'Etat, partie civile dans l'affaire, a été empêchée d'assister à l'audience. https://www.lorientlejour.com/article/1426489/riad-salame-est-arrive-au-palais-de-justice-sous-les-huees-des-manifestants.html?utm_source=olj&%E2%80%A6

Orhan, Z. (2023, Kasım 7). *Küresel İslami Finans Raporundan Son Durum Notları, 2023*. https://islamiktisadi.net: https://islamiktisadi.net/2023/11/07/kuresel-islami-finans-raporundan-son-durum-notlari-2023/

Oseni, U., & Ali, S. (2019). *Fintech in Islamic Finance: Theory and Practice*. Routledge. DOI: 10.4324/9781351025584

Özsoy, Ş. (2012). *Sağlam bankacılık modeli ile katılım bankacılığına giriş*. Kuveyt Türk Yayınları.

Penn, A., & Arias, M. (2009). *Global E-Business Law & Taxation*. Oxford University Press. DOI: 10.1093/oso/9780195367218.001.0001

Pietro Biancone, P., Secinaro, S., & Kamal, M. (2019). Crowdfunding and Fintech: Business model sharia compliant. *European Journal of Islamic Finance*, *12*, 1–10. DOI: 10.13135/2421-2172/3260

Pireh, M. (2008). *Definition of Islamic Financial Instruments, Islamic Finance Expert Securites & Exchange Organization (SEO), 21*. Tahran.

Polzin, F., Egli, F., Steffen, B., & Schmidt, T. S. (2019). How do policies mobilize private finance for renewable energy?—A systematic review with an investor perspective. *Applied Energy*, *236*, 1249–1268. DOI: 10.1016/j.apenergy.2018.11.098

Porretta, P., & Benassi, A. (2021). Sustainable vs. not sustainable cooperative banks business model: The case of GBCI and the authority view. *Risk Governance and Control: Financial Markets & Institutions*, *11*(1), 33–48. DOI: 10.22495/rgcv11i1p3

PwC. (2024, June 5). *Sustainability in the Middle East 2024*. PwC. https://www.pwc.com/m1/en/sustainability/insights/sustainability-in-the-middle-east-2024.html

Qudah, H., Malahim, S., Airout, R. M., Alomari, M., Hamour, A. A., & Alqudah, M. (2023). Islamic Finance in the Era of Financial Technology: A Bibliometric Review of Future Trends. *International Journal of Financial Studies*, *11*(2), 1–29. DOI: 10.3390/ijfs11020076

Rabbani, M. R., Sarea, A., Khan, S., & Abdullah, Y. (2022b). Ethical concerns in artificial intelligence (AI): The role of RegTech and Islamic finance. In Artificial Intelligence for Sustainable Finance and Sustainable Technology: Proceedings of ICGER 2021 1. Cham: Springer International Publishing, pp. 381–390.

Rabbani, M., Khan, S., & Thalassinos, E. (2020). FinTech, Blockchain and Islamic Finance: An Extensive Literature Review. *International Journal of Economics and Business Administration*, 65-86.

Rabbani, M. R. (2022a). Fintech innovations, scope, challenges, and implications in Islamic Finance: A systematic analysis. *International Journal of Computing and Digital Systems*, *11*, 579–608.

Rahman, M. H., Rahman, J., Tanchangya, T., & Esquivias, M. A. (2023). Green banking initiatives and sustainability: A comparative analysis between Bangladesh and India. *Research in Globalization*, *7*, 100184. DOI: 10.1016/j.resglo.2023.100184

Rahman, S., & Ali, M. (2021). *Strategic Change*, *30*(3), 261–275.

Raimi, L., Abdur-Rauf, I. A., & Ashafa, S. A. (2024). Does Islamic sustainable finance support sustainable development goals to avert financial risk in the management of Islamic finance products? A critical literature review. *Journal of Risk and Financial Management*, *17*(6), 236. DOI: 10.3390/jrfm17060236

Rani, A. M., Dariah, A. R., Al Madhoun, W., & Srisusilawati, P. (2023). Awareness of sustainable finance development in the world from a stakeholder perspective. *International Journal of Management and Sustainability*, *12*(3), 323–336. DOI: 10.18488/11.v12i3.3428

Raza, A., Alavi, A. B., & Asif, L. (2024). Sustainability and financial performance in the banking industry of the United Arab Emirates. *Discover Sustainability*, *5*(1), 223. DOI: 10.1007/s43621-024-00414-z

Razak, D. A., Zulmi, S. R., & Dawami, Q. (2021). Customers' perception on islamic crowdfunding as a possible financial solution for the pandemic COVID-19 crisis in Malaysia. *Journal of Islamic Finance*, *10*, 92–100.

Raza, S., & Ding, C. (2021). News recommender system: A review of recent progress, challenges, and opportunities. *Artificial Intelligence Review*, *55*(1), 749–800. DOI: 10.1007/s10462-021-10043-x PMID: 34305252

Reynolds, S. (2024). Stakeholder engagement and its impact on supply chain sustainability in the context of renewable energy. *Preprints*. https://doi.org/ DOI: 10.20944/preprints202406.0080.v1

Rizal, F., & Rofiqo, A. (2020). Determinants of Sharia Banking Profitability: Empirical Studies in Indonesia 2011-2020. *El Barka: Jouranl of Islamic Economic and BUsiness*, *3*(1), 137–161. DOI: 10.21154/elbarka.v3i1.2051

Ross, M. (2020). Working with Islamic Finance. investopedia. https://www.investopedia.com/articles/07/islamic_investing.asp

Rosyadah, P. C., Arifin, N. R., Muhtadi, R., & Safik, M. (2020). Factors That Affect Savings In Islamic Banking. *AL-ARBAH:Journal of Islamic Finance and Banking*, *2*(1), 33–46.

Roy, D., & Dutta, M. (2022). A systematic review and research perspective on recommender systems. *Journal of Big Data*, *9*(1), 59. DOI: 10.1186/s40537-022-00592-5

Saba, I., Kouser, R., & Chaudhry, I. S. (2019). FinTech and Islamic finance-challenges and opportunities. *Review of Economics and Development Studies*, *5*(4), 581–890. DOI: 10.26710/reads.v5i4.887

Samar, M., & Şimşek, M. (2020). İslami Finans Açısından Blokzincir Teknolojisi. *Necmettin Erbakan Üniversitesi Yayınları*, 81-104.

Saradara, S. M., Khalfan, M. M. A., Rauf, A., & Qureshi, R. (2023). On the path towards sustainable construction—the case of the United Arab Emirates: A review. *Sustainability (Basel)*, *15*(19), 14652. DOI: 10.3390/su151914652

Sarath, K. N. V. D., & Ravi, V. (2013). Association rule mining using binary particle swarm optimization. *Engineering Applications of Artificial Intelligence*, *26*(8), 1832–1840. DOI: 10.1016/j.engappai.2013.06.003

Selcuk, M., & Kaya, S. (2021). A Critical Analysis of Cryptocurrencies from an Islamic Jurisprudence Perspective. *Turkish Journal Of Islamic Economics*, *8*(1), 137–152. DOI: 10.26414/A130

Shafii, Z., Zakaria, N., Sairally, B. S., Shaharuddin, A., Hussain, L., & Zuki, M. S. M. (2013). *An Appraisal of the Principles Underlying International Financial Reporting Standards (IFRS): A Shari'ah Perspective-Part 1*. International Shari'ah Research Academy for Islamic Finance (ISRA).

Shalev-Shwartz, S., & Ben-David, S. (2014). *Understanding Machine Learning: From Theory to Algorithms*. Cambridge University Press. DOI: 10.1017/CBO9781107298019

iddīqī, M. N. (1981). Muslim economic thinking: A survey of contemporary literature. Islamic economics series.

Siddiqi, M. (1982). *Nejatullah*. Islamic Approaches to Money, Banking and Monetary Policy.

Singh, R., & Kaur, H. (2022). *Journal of Islamic Business and Management*, *12*(1), 89–105.

Siska, E. (2022). Exploring the Essential Factors on Digital Islamic Banking Adoption in Indonesia: A Literature Review. *Jurnal Ilmiah Ekonomi Islam, 8*(1), 124–130. DOI: 10.29040/jiei.v8i1.4090

Siska, E., Gamal, A. A. M., Ameen, A., & Amalia, M. M. (2021). Analysis Impact of Covid-19 Outbreak on Performance of Commercial Conventional Banks: Evidence from Indonesia. *International Journal of Social and Management Studies*, *2*(6), 8–16.

Spendolini, M. J. (2021). *The Benchmarking Book*. American Management Association.

Srivastava, T., Mullick, I., & Bedi, J. (2024). Association mining based deep learning approach for financial time-series forecasting. *Applied Soft Computing*, *155*, 111469. DOI: 10.1016/j.asoc.2024.111469

Srour, I., & Chaaban, J. (2017). *Market Research Study on the Development of Viable Economic Subsectors in Lebanon*. EU Project.

Statista. (2024). *Revenue of fintech industry worldwide 2017-2028*. https://www.statista.com/statistics/1384016/estimated-revenue-of-global-fintech/

Stempel, J. (2019). UBS must defend against U.S. lawsuit over 'catastrophic' mortgage losses. Yahoo Finance. Retrieved from https://finance.yahoo.com/news/ubs-must-defend-against-u -214743943.html

Stupples, B., Sazonov, A., & Woolley, S. (2019, July 26). UBS Whistle-Blower Hunts Trillions Hidden in Treasure Isles. Bloomberg. Retrieved from https://www.bloomberg.com/news/articles/2019-07-26/ubs-whistle-blower-hunts-trillions-hidden-in-treasure-islands

Sujud, H., & Hachem, B. (2018). Reality and future of islamic banking in lebanon. *European Journal of Scientific Research*.

Taghizadeh-Hesary, F., & Yoshino, N. (2020). Sustainable solutions for green financing and investment in renewable energy projects. *Energies*, *13*(4), 788. DOI: 10.3390/en13040788

The Open Group. (2011a). *The Open Group's Architecture Framework*.

The Open Group. (2011b). *Architecture Development Method*. The Open Group.

The World Bank Group. (2020). Mobilizing private finance for nature. Retrieved from https://thedocs.worldbank.org/en/doc/916781601304630850-0120022020/original/FinanceforNature28Sepwebversion.pdf

Trad, A., & Kalpić, D. (2018c). The Business Transformation Framework and Enterprise Architecture Framework for Managers in Business Innovation. The role of legacy processes in automated business environments. The Proceedings of E-LEADER 2017 Berlin, 1.

Trad, A., & Kalpić, D. (2019a). The Business Transformation Framework and the Application of a Holistic Strategic Security Concept. Chinese American Scholars Association Conference E-Leader, Conference, Brno. Check Republic.

Trad, A. (2018e). *The Business Transformation and Enterprise Architecture Framework Applied to analyse-The historically recent Rise and the 1975 Fall of the Lebanese Business Ecosystem*. IGI-Global.

Trad, A., & Kalpić, D. (2017b). *A Neural Networks Portable and Agnostic Implementation IHIPTF for Business Transformation Projects. The Basic Structure*. IEEE.

Trad, A., & Kalpić, D. (2017c). *A Neural Networks Portable and Agnostic Implementation IHIPTF for Business Transformation Projects. The Framework*. IEEE.

Trad, A., & Kalpić, D. (2018a). *The Business Transformation Framework and Enterprise Architecture Framework for Managers in Business Innovation-Knowledge Management in Global Software Engineering (HKMS)*. IGI-Global.

Trad, A., & Kalpić, D. (2018f). *An applied mathematical model for business transformation-The Holistic Critical Success Factors Management System (HCSFMS). Encyclopaedia of E-Commerce Development, Implementation, and Management*. IGI-Global.

Trad, A., & Kalpić, D. (2020a). *Using Applied Mathematical Models for Business Transformation. Author Book*. IGI-Global. DOI: 10.4018/978-1-7998-1009-4

Trad, A., Nakitende, M., & Oke, T. (2021). *Tech-Based Enterprise Control and Audit for Financial Crimes: The Case of State-Owned Global Financial Predators (SOGFP). Book: Handbook of Research on Theory and Practice of Financial Crimes*. IGI Global.

Truby, J., & Ismailov, O. (2022). The role and potential of blockchain technology in Islamic finance. *European Business Law Review*, *33*(2), 175–192. DOI: 10.54648/EULR2022005

Turan, M. (2021). *Türkiye'de İslami Fintechlerin Güncel Durumu.* islamiktisadi.net: https://islamiktisadi.net/2021/02/23/turkiyede-islami-fintechlerin-guncel-durumu/

Ukoba, K., Yoro, K. O., Eterigho-Ikelegbe, O., Ibegbulam, C., & Jen, T.-C. (2024). Adaptation of solar energy in the Global South: Prospects, challenges and opportunities. *Heliyon*, *10*(7), e28009. DOI: 10.1016/j.heliyon.2024.e28009 PMID: 38560131

UN. (2020). Lebanon: UN-backed tribunal sentences Hezbollah militant in Hariri assassination. UN. https://news.un.org/en/story/2020/12/1079892

UN. (2023). Justice served: Lebanon's Special Tribunal closes. UN. https://news.un.org/en/story/2023/12/1145217

UNEP FI. (2020). Draft for consultation: Promoting sustainable finance & climate finance in the Arab region. United Nations Environment Programme Finance Initiative. https://www.unepfi.org/wordpress/wp-content/uploads/2020/10/ConsultationDraft-Promoting-Sustainable-Finance-in-the-Arab-Region.pdf

Unger, C., & Thielges, S. (2021). Preparing the playing field: Climate club governance of the G20, Climate and Clean Air Coalition, and Under2 Coalition. *Climatic Change*, *167*(3-4), 41. DOI: 10.1007/s10584-021-03189-8

United Nations. (2018). Towards Saudi Arabia's sustainable tomorrow: Kingdom of Saudi Arabia, UN High-Level Political Forum 2018: "Transformation towards sustainable and resilient societies". https://sustainabledevelopment.un.org/content/documents/20230SDGs_English_Report972018_FINAL.pdf

United Nations. (2023). The state of Arab cities report 2022. United Nations Development Programme (UNDP). https://www.undp.org/sites/g/files/zskgke326/files/2024-05/the_state_of_arab_cities_report_2022_final_eng.pdf

United Nations. (2024). The Sustainable Development Goals Report 2024. United Nations. https://unstats.un.org/sdgs/report/2024/The-Sustainable-Development-Goals-Report-2024.pdf

Ünlü, R. (2022). An Assessment of imbalanced control chart pattern recognition by artificial neural networks. İçinde Research Anthology on Artificial Neural Network Applications (ss. 683-702). IGI Global. https://www.igi-global.com/chapter/an-assessment-of-imbalanced-control-chart-pattern-recognition-by-artificial-neural-networks/288982

Ünlü, R., & Xanthopoulos, P. (2021). A reduced variance unsupervised ensemble learning algorithm based on modern portfolio theory. *Expert Systems with Applications*, *180*, 115085. DOI: 10.1016/j.eswa.2021.115085

Vassileva, A. (2022). Green public-private partnerships (PPPs) as an instrument for sustainable development. *Journal of World Economy: Transformations & Transitions*, *2*(5), 1–18. DOI: 10.52459/jowett25221122

Wang, J., Liu, Z., & Yang, X. (2021). *Journal of Quantitative Analysis in Finance*, *19*(2), 123–140.

Weber, O. (2018). The financial sector and the SDGs: Interconnections and future directions (CIGI Book chapters No. 201). Centre for International Governance Innovation. https://www.cigionline.org/static/documents/documents/Book chapter%20No.201web.pdf

Wikipedia. (2020a). Islamic banking and finance. Wikipedia, the free encyclopedia. https://en.wikipedia.org/wiki/Islamic_banking_and_finance

Wikipedia. (2024a). Raymond Eddé. https://en.wikipedia.org/wiki/Raymond_Edd%C3%A9

Witten, I. H., Frank, E., Hall, M. A., & Pal, C. J. (2017). *Data Mining: Practical Machine Learning Tools and Techniques*. Morgan Kaufmann.

World Bank Group. (2017, March). Morocco: Noor Ouarzazate - Concentrated solar power complex. Multilateral Development Banks' Collaboration: Infrastructure Investment Project Briefs. https://ppp.worldbank.org/public-private-partnership/sites/ppp.worldbank.org/files/2022-02/MoroccoNoorQuarzazateSolar_WBG_AfDB_EIB.pdf

Yıldız, N. (2023). Türkiye'de Faaliyet Gösteren Katılım Bankalarının Performans Analizi. *Cumhuriyet Üniversitesi İktisadi ve İdari Bilimler Dergisi*, *24*(1), 36–49. DOI: 10.37880/cumuiibf.1173166

Yılmaz, C., & Özgür, E. (2021). Katılım Bankalarında Kârlılığa Etki Eden Faktörlerin Tespiti İçin Panel Veri Analizi Uygulaması. *Muhasebe Ve Finansman Dergisi*, (92), 1–20. DOI: 10.25095/mufad.944461

Yumurtacı, R. (2023). Türkiye'deki Konvansiyonel Bankalar ile Katılım Bankalarının CAMELS Analizi ile Karşılaştırılması. *Süleyman Demirel Üniversitesi Vizyoner Dergisi*, *14*(39), 1077–1097. DOI: 10.21076/vizyoner.1196650

Zaidan, E., Al-Saidi, M., & Hammad, S. H. (2019). Sustainable development in the Arab world – Is the Gulf Cooperation Council (GCC) region fit for the challenge? *Development in Practice*, *29*(5), 670–681. DOI: 10.1080/09614524.2019.1628922

Zalloua, P. (2004). The NGM Study "Who were the SPs" and the Return of the SPs. Interview Lebanese Broadcasting Corporation (LBC). Retrieved April 16, 2017, from. Phoenicia. National Geographic (October 2004).

Zamiri, M., & Esmaeili, A. (2024). Methods and technologies for supporting knowledge sharing within learning communities: A systematic literature review. *Administrative Sciences*, *14*(1), 17. DOI: 10.3390/admsci14010017

Zhang, L., Wang, Y., & Liu, Q. (2022). *Emerging Markets Finance & Trade*, *58*(5), 1342–1360.

Zhou, K., Luo, L., & Chen, T. (2023). The Influence of Online Media on College Students' Self-identity in Mobile Learning Environment. In 2023 4th International Conference on Education, Knowledge and Information Management (ICEKIM 2023), 1195-1203.

Zhou, L., & Yau, S. (2007). Efficient association rule mining among both frequent and infrequent items. *Computers & Mathematics with Applications (Oxford, England)*, *54*(5), 737–749. DOI: 10.1016/j.camwa.2007.02.010

Zhou, Y., & Wu, H. (2022). Journal of Islamic Economics. *Banking and Finance*, *18*(1), 67–85.

Zuhroh, I. (2021). The impact of Fintech on Islamic banking and the collaboration model: A systematic review studies in Indonesia. *Jurnal Perspektif Pembiayaan Dan Pembangunan Daerah*, *9*(4), 301–312. DOI: 10.22437/ppd.v9i4.12054

Zulkhibri, M. (2019). Fintech and the Future of Islamic Finance: Opportunities and Challenges. *Journal of Islamic Monetary Economics and Finance*, *5*, 629–652.

About the Contributors

Yilmaz Bayar awarded the PhD degree in the field of Economics in Istanbul University Institute of Social Sciences in 2012. He worked as an Assistant Professor in the Faculty of Business Administration, Karabuk University during the period September 2012-March 2015 and as an Associate Professor in the Faculty of Economics and Administrative Sciences, Usak University, during the period March 2015-October 2020 and now working as a Professor in the Faculty of Economics and Administrative Sciences, Bandirma Onyedi Eylul University, since October 2020.

Mohamed Sadok Gassouma is an associate Professor have an hability to conduct research. I obtained a doctoral degree then an habilitation diploma in finance delivered by the Faculty of Economics and Management of Tunis in Tunisia. I taught from 2006 until 2009 as a contractual assistant in finance at the higher business school of Tunis then as a permanent assistant from 2010 until 2012 at the higher institute of Finance and Taxation of Sousse as an assistant professor from 2013 until 2021 and from 2022 until now a as an associate Professor having an hability to conduct research. I started my teachings in Islamic finance at 2018 until now at the Higher Institute of Theology University EZ-Zitouna.

***

Burak Aktürk completed his undergraduate education at Marmara University, Department of Economics in 2017. He received his master's degree from Marmara University, Department of Islamic Finance and Banking in 2021 with his thesis titled "Financial Technologies in Islamic Finance and Fintechs' Participation Banking Applications". He has worked in various units at Kuveyt Türk Participation Bank and is currently continuing his professional career at Kuveyt Türk Participation Bank. He is interested in participation finance and Islamic fintech and has published various articles on these topics.

Raed Awashreh is Professor of Management & Public Administration. He brings over 25 years of combined academic and industry experience, teaching undergraduate and postgraduate courses. Dr. Awashreh has authored 50 articles, several Book chapters, and 3 books spanning topics such as, human resources, governance, public policy, strategy, leadership, Entrepreneurship, Innovation, multidisciplinary linked AI, and management. In addition to his academic role, Dr. Awashreh serves as a consultant in business, management, governance, HR, and organizational development, working closely with government, private, and non-governmental sectors. He earned his degrees in Public Administration and Management from Flinders University, Australia, and Monterey, USA.

Muhammet Enis Bulak was born in 1986 in Istanbul. He completed his Ph.D. in 2018 in the Department of Industrial Engineering at Istanbul University. In 2024, he was awarded the title of Associate Professor in the field of Industrial Engineering. Focusing on sustainability, production management, and project management—some of the most pressing and contemporary topics of today—Bulak also conducts research in innovation, quality management, multivariate decision-making, product design, and usability. In 2019, he served as a visiting faculty member at Qatar University. His writings and interviews on circular economy, resource efficiency, environmental awareness, quality control, branding, and Turquality have been featured in various media outlets. Bulak continues his research on Green Supply Chain Management, a critical emerging need for the production and service sectors.

Mohammad Chebli, Industrial Engineer skilled in blending quantitative analysis with practical decision-making to enhance organizational performance. Through experience in feasibility assessments, data modeling, and simulation-driven evaluations, He has supported complex projects by identifying efficiency gaps, forecasting trends, and streamlining resource utilization. Adept at working with a range of analytical tools—from advanced Excel automation to computational modeling platforms—He excels at turning detailed data into strategic insights that inform long-term growth and operational effectiveness.

Nizamülmülk Güneş received a bachelor's degree in Economics at Inonu University in 1997. He finished an MA in human resource at Istanbul University in 2004. Then, he finished doctorate in the field of economics and finance at Kocaeli University in 2014. In between the period 1997-2001, he worked as an auditor some private banks. Nizamülmülk Güneş is an auditor at the Savings Deposit Insurance Fund in Istanbul between 2001-2023. He has been working Marmara University since 2023. He had been interested in finance, audit and macroeconomics.

Yunus Emre Gürbüz was born in Bayburt in 1979. After he has graduated from Istanbul University in Department of Economics he has worked in different sectors like banking, reel and insurance sectors as an audit assistant and then auditor since 2004. He has been working in Neova Sigorta which is takaful company as a Head of Internal Audit, Takaful Association Secretary General since 2010. Meanwhile he has placed among the founders of Takaful Association which is known as first nongovernmental organization in the name of Takaful Association. He is also Secretary General of Takaful Association.

Funda Hatice Sezgin completed her undergraduate education in Faculty of economics in University of Istanbul. She did master of business administration degree (MBA) in Econometrics of University of Istanbul, and doctor of econometrics in Marmara University. Her research interests are time series analysis, panel data analysis and statistical methods. She works in the department of industrial engineering at Istanbul University-Cerrahpasa.

Yusuf Sait Turkan is working as Associate Professor in the Department of Industrial Engineering of Istanbul University - Cerrahpasa, Turkey. He is a also the coordinator of Distance Industrial Engineering Program at

Istanbul University Open and Distance Faculty. He has completed his PhD and Master of Science in Industrial Engineering at the Istanbul University. His research interest include system simulation, metaheuristic algorithms and machine learning applications.

Index

A

B

C

D

E

F

G

I

L

M

O

P

S

T

Printed in the United States
by Baker & Taylor Publisher Services